D1105995

INTERNATIONAL UNION OF PURE AND APPLIED CHEMISTRY

NOMENCLATURE OF INORGANIC CHEMISTRY

1970

IUPAC COMMISSION ON

THE NOMENCLATURE OF INORGANIC CHEMISTRY

The membership of the Commission during the period 1959–1971 in which the present edition was prepared was as follows:

Titular Members

Chairman: 1959–1971 K. A. JENSEN (Denmark).

Honorary Chairmen: 1959–1965 H. BASSETT† (U.K.); 1959–1961 A. SILVERMAN† (U.S.A.).

Vice-Chairman: 1959–1971 H. REMY (Germany).

Secretaries: 1959–1963 J. CHATT (U.K.); 1959–1971 F. GALLAIS (France); 1963–1971 J. E. PRUE (U.K.).

Members: 1967–1971 R. M. ADAMS (U.S.A.); 1959–1963 J. BÉNARD (France); 1963–1971 J. CHATT (U.K.); 1959–1971 G. H. CHEESMAN (Australia); 1959–1967 E. J. CRANE† (U.S.A.); 1963–1971 W. C. FERNELIUS (U.S.A.); 1959–1963 W. FEITKNECHT (Switzerland); 1959–1971 L. MALATESTA (Italy); 1959–1971 A. ÖLANDER (Sweden).

Associate Members

1963–1967 J. BÉNARD (France); 1969–1971 L. F. BERTELLO (Argentina); 1969–1971 K.-C. BUSCHBECK (Germany); 1969–1971 T. ERDEY-GRÚZ (Hungary); 1963–1967 W. FEITKNECHT (Switzerland); 1959–1963 W. C. FERNELIUS (U.S.A.); 1969–1971 Y. JEANNIN (France); 1959–1965 W. KLEMM (Germany); 1959–1967 W. KOTOWSKI (Germany); 1969–1971 W. H. POWELL (U.S.A.); 1963–1971 A. L. G. REES (Australia); 1969–1971 A. A. VLČEK (Czechoslovakia); 1969–1971 E. WEISS (Germany); 1959–1971 K. YAMASAKI (Japan).

Observers

From Commission III.1: 1959–1971 S. VEIBEL (Denmark).

From Commission IV.1: 1967–1971 K. L. LOENING (U.S.A.).

†Deceased

INTERNATIONAL UNION OF
PURE AND APPLIED CHEMISTRY

NOMENCLATURE OF INORGANIC CHEMISTRY

SECOND EDITION

DEFINITIVE RULES 1970

Issued by the Commission on the Nomenclature of Inorganic Chemistry

LONDON
BUTTERWORTHS

THE BUTTERWORTH GROUP

ENGLAND:	BUTTERWORTH & CO. (PUBLISHERS) LTD. LONDON: 88 Kingsway, WC2B 6AB
AUSTRALIA:	BUTTERWORTH & CO. (AUSTRALIA) LTD. SYDNEY: 20 Loftus Street MELBOURNE: 343 Little Collins Street BRISBANE: 240 Queen Street
CANADA:	BUTTERWORTH & CO. (CANADA) LTD. TORONTO: 14 Curity Avenue, 374
NEW ZEALAND:	BUTTERWORTH & CO. (NEW ZEALAND) LTD. WELLINGTON: 49/51 Ballance Street AUCKLAND: 35 High Street
SOUTH AFRICA:	BUTTERWORTH & CO. (SOUTH AFRICA) (PTY) LTD. DURBAN: 33–35 Beach Grove

First Edition (*1957 Rules*) published in 1959

Second Edition (*1970 Rules*) published in 1971

The contents of this book appear in

Pure and Applied Chemistry, Vol. 28. No. 1 (1971)

Suggested U.D.C. *number*: 547·9(063)

ISBN: 0 408 70168 4

Printed in Great Britain by Page Bros. (Norwich) Ltd., Norwich

PREFACE TO THE FIRST EDITION

IN addition to members of the Commission on Inorganic Chemical Nomenclature listed in the footnote on page 1, the present revision is the evolved work of various individuals who have served as regular members of the Commission since the "1940 Rules" appeared. Their names are listed in the volumes of Comptes Rendus, I.U.P.A.C., which have appeared since 1940.

Acknowledgement is also made of the co-operation of delegate and advisory members of the Commission, of members of nomenclature committees in a number of nations; also of Dr E. J. Crane, Editor of *Chemical Abstracts*.

The final editing of the 1957 Report is the work of a sub-committee, Professor K. A. Jensen, Chairman, Professor J. Bénard, Professor A. Ölander and Professor H. Remy.

Alexander Silverman
Chairman

November 1st, 1958

PREFACE TO THE SECOND EDITION

The IUPAC Commission on the Nomenclature of Inorganic Chemistry, in its first meeting after the publication of the 1957 Rules (Munich 1959), scheduled further work for the Commission to deal with the nomenclature of boron hydrides and higher hydrides of the Group IV–VI elements, polyacids, and organometallic compounds. Part of this work, dealing with organometallic compounds, organoboron, organosilicon, and organophosphorus compounds, has been carried out in collaboration with the IUPAC Commission on the Nomenclature of Organic Chemistry. It has now been completed and Tentative Rules for this field will be published.

In the meantime work on a revision of the 1957 Rules has been going on continuously. Tentative proposals for changes or additions to this Report have been published in the Comptes Rendus of the Conferences in London (1963) and Paris (1965) and in the *IUPAC Information Bulletin*. However, as a result of comments to these tentative proposals some of them (such as the proposal to change chloro to chlorido) have not been retained in the final version. The section on coordination compounds has been much extended, reflecting the importance of this field in modern inorganic chemistry. A short section on boron hydrides and their derivatives has been included in the present edition, but a more extensive treatment has been published separately on a tentative basis in the *IUPAC Information Bulletin:* Appendices on Tentative Nomenclature, Symbols, Units and Standards, No. 8 (September 1970).

An editorial committee consisting of Prof. R. M. Adams, Prof. J. Chatt, Prof. W. C. Fernelius, Prof. F. Gallais, Dr. W. H. Powell, and Dr. J. E. Prue met in Columbus, Ohio, in the last week of January 1970, to finalize the manuscript. The Commission acknowledges the help of Dr. Kurt Loening, Chairman of the Commission on the Nomenclature of Macromolecular Chemistry, during this work. The work of the Commission was aided significantly by Grant No. 890–65 from the Air Force Office of Scientific Research administered by the U.S.A. National Academy of Sciences – National Research Council.

K. A. JENSEN
Chairman
Commission on the Nomenclature
of Inorganic Chemistry

Copenhagen
September 9th, 1970

CONTENTS

Page

PREFACE TO THE FIRST EDITION v

PREFACE TO THE SECOND EDITION vii

INTRODUCTION TO THE FIRST EDITION 1

INTRODUCTION TO THE SECOND EDITION 3

0. PREAMBLE 5
   - 0.1 Oxidation Number 5
   - 0.2 Coordination Number 6
   - 0.3 Use of Multiplying Affixes, Enclosing Marks, Numbers and Italic Letters 6
     - 0.31 Multiplying affixes 6
     - 0.32 Enclosing marks 7
     - 0.33 Numbers 8
     - 0.34 Italic letters 9

1. ELEMENTS 10
   - 1.1 Names and Symbols of the Elements 10
   - 1.2 Names for Groups of Elements and their Subdivisions 10
   - 1.3 Indication of Mass, Charge, *etc.*, on Atomic Symbols .. 11
   - 1.4 Allotropes 12

2. FORMULAE AND NAMES OF COMPOUNDS IN GENERAL 13
   - 2.1 Formulae 13
   - 2.2 Systematic Names 15
   - 2.3 Hydrides 18
   - 2.4 Trivial Names 19

3. NAMES FOR IONS AND RADICALS 20
   - 3.1 Cations 20
   - 3.2 Anions 21
   - 3.3 Radicals 23

4. ISO- AND HETEROPOLYANIONS 26
   - 4.1 Isopolyanions 26
   - 4.2 Heteropolyanions 28

5. ACIDS 31
5.1 BINARY AND PSEUDOBINARY ACIDS 31
5.2 ACIDS DERIVED FROM POLYATOMIC ANIONS 31
5.214 Names for oxoacids 32
5.3 FUNCTIONAL DERIVATIVES OF ACIDS 34

6. SALTS AND SALT-LIKE COMPOUNDS 36
6.1 SIMPLE SALTS 36
6.2 SALTS CONTAINING ACID HYDROGEN ("Acid" salts) .. 36
6.3 DOUBLE, TRIPLE, *etc.*, SALTS 36
6.4 OXIDE AND HYDROXIDE SALTS ("Basic" salts) 37
6.5 DOUBLE OXIDES AND HYDROXIDES 38

7. COORDINATION COMPOUNDS 39
7.1 DEFINITIONS 39
7.2 FORMULAE AND NAMES FOR COORDINATION COMPOUNDS IN GENERAL 39
7.21 Central atoms 39
7.22 Indication of oxidation number and proportion of constituents 40
7.23 Structural prefixes 40
7.24 Terminations 40
7.25 Order of citation of ligands in coordination entities 41
7.3 NAMES FOR LIGANDS 41
7.31 Anionic ligands 41
7.32 Neutral and cationic ligands 44
7.33 Different modes of linkage of some ligands .. 46
7.34 Designation of active coordination sites 47
7.35 Use of abbreviations 47
7.4 COMPLEXES WITH UNSATURATED MOLECULES OR GROUPS 49
7.41 Designation of stoicheiometric composition only 49
7.42 Designation of structure 49
7.43 Cyclopentadienyl complexes: Metallocenes .. 52
7.5 DESIGNATION OF ISOMERS 55
7.51 Geometrical isomerism 56
7.52 Isomerism due to chirality (asymmetry) .. 64
7.6 DI- AND POLYNUCLEAR COMPOUNDS WITH BRIDGING GROUPS 65
7.61 Compounds with bridging atoms or groups .. 65

7.62 Extended structures 71

7.7 DI- AND POLYNUCLEAR COMPOUNDS WITHOUT BRIDGING GROUPS 72

7.71 Direct linking between centres of coordination .. 72

7.72 Homoatomic aggregates 73

7.8 ABSOLUTE CONFIGURATIONS CONCERNED WITH SIX-COORDINATED COMPLEXES BASED ON THE OCTAHEDRON .. 75

7.81 Basic principle 77

7.82 Application to configuration 77

7.83 Application to conformation 80

7.84 Absolute configurations 81

7.85 Phenomenological characterization 81

7.86 Full characterization 81

7.87 Designation of configurational chirality caused by chelation in six-coordinated complexes based on the octahedron 81

7.88 Designation of conformational chirality of a chelate ring 82

Appendix—Relationships between the proposed symbols and those in earlier use 82

8. ADDITION COMPOUNDS 84

9. CRYSTALLINE PHASES OF VARIABLE COMPOSITION .. 86

10. POLYMORPHISM 90

11. BORON COMPOUNDS 92

11.1 BORON HYDRIDES 92

11.2 BORANES WITH SKELETAL REPLACEMENT 95

11.3 BORON RADICALS 96

11.4 SUBSTITUTION PRODUCTS OF BORANES 96

11.5 ANIONS DERIVED FROM THE BORANES 97

11.6 CATIONS DERIVED FROM THE BORANES 97

11.7 SALTS DERIVED FROM THE BORANES 97

TABLE I ELEMENTS 98

TABLE II NAMES FOR IONS AND RADICALS 99

TABLE III PREFIXES OR AFFIXES USED IN INORGANIC NOMENCLATURE 103

TABLE IV ELEMENT SEQUENCE 104

TABLE V ELEMENT RADICAL NAMES 105

INDEX 107

INTRODUCTION TO THE FIRST EDITION

(1957 RULES)

The Commission on the Nomenclature of Inorganic Chemistry of the International Union of Pure and Applied Chemistry (I.U.P.A.C.) was formed in 1921, and many meetings took place which culminated in the drafting of a comprehensive set of Rules in 1938. On account of the war they were published in 1940 without outside discussion. At the meeting of the International Union of Chemistry in 1947 it was decided to undertake a thorough revision of what have come to be known as the "1940 Rules", and after much discussion they were completely rewritten and issued in English and French, the official languages of the Union, after the meeting in Stockholm in 1953 as the "Tentative Rules for Inorganic Chemical Nomenclature". These were studied by the various National Organizations and the comments and criticisms of many bodies and of private individuals were received and considered in Zürich, Switzerland, in 1955, in Reading, England, in 1956, and in Paris, France, in 1957.

The Rules set out here express the opinion of the Commission* as to the best general system of nomenclature, although certain names and rules which are given here as a basis for uniformity will probably prove unworkable or unacceptable in some languages. In these cases adaptation or even alteration will be necessary, but it is hoped that it will be possible to keep these changes small and to preserve the spirit of the I.U.P.A.C. Rules. The English and French versions, which differ slightly, are to be regarded as international models from which translations will be made into other languages. The latter is likely to prove the better model for the Romance languages, and the former for Germanic languages, but it must be borne in mind that these languages are here used as the official languages of the Union and that several nations speak them with quite considerable variations of usage among themselves. There may therefore arise a similar need for adaptation or alteration even among English-speaking and French-speaking peoples, but we hope that in these cases, as in the others, careful consideration will be given to minimizing variation and to preserving the spirit of our international model.

The Commission's aim has been always to produce rules which lead to clear and acceptable names for as many inorganic compounds as possible. It soon became obvious, however, that different users may require the name of a compound to fulfil different objects, and it has been necessary to effect compromises in order to formulate rules of the most general utility. The principal function of a name is to provide the chemist with a word or set of words which is unique to the particular compound, and which conveys at least its empirical formula and also if possible its principal structural

* Chairman (1947–53) H. Bassett; (1953–57) Alex Silverman; Vice-Chairman, K. A. Jensen; Secretary, G. H. Cheesman; Members, J. Bénard, N. Bjerrum, E. H. Büchner, W. Feitknecht, L. Malatesta, A. Ölander, and H. Remy.

features. The name should be pronounceable and capable of being written or printed with an absolute minimum of additional symbols or modes of writing (*e.g.*, subscripts or differing type-faces).

Many inorganic compounds exist only in the solid state, and are destroyed on fusion, solution or vaporization; some chemists have expressed strongly the view that names for such compounds should include information about the structure of the solid as well as its composition. Incorporating all this information tends to make the names extremely cumbersome, and since many of the structures remain uncertain or controversial, the names themselves are apt to be unstable. For general purposes, therefore, a considerable curtailment is essential and the Commission has endeavoured to produce a system based on the composition and most obvious properties of substances, avoiding as far as possible theoretical matters which are liable to change.

INTRODUCTION TO THE SECOND EDITION

(1970 RULES)

A major revision and extension of Section 7 has been undertaken. The principle of an alphabetical order of citation of ligands in coordination entities has been adopted, and the rules now make detailed provision for the naming of complexes with unsaturated molecules or groups, the designation of ligand positions in the coordination sphere, the nomenclature of polynuclear compounds and those with metal-metal bonds, and the nomenclature of absolute configurations for six-coordinated complexes based on the octahedron. The former Section 4 which dealt with crystalline phases of variable composition has been similarly revised and extended, and now becomes Section 9. Its place as Section 4 is taken by a fuller treatment of polyanions, formerly briefly dealt with in a sub-section of Section 7. The rules for the nomenclature of inorganic boron compounds are outlined in Section 11. Extended tentative rules will be found in the *IUPAC Information Bulletin:* Appendices on Tentative Nomenclature, Symbols, Units and Standards, No. 8 (September 1970). The alphabetical principle, already mentioned in connection with Section 7, has also now been widely adopted in Sections 2 and 6.

The introduction of a preamble will, it is hoped, make clear more readily than is possible within the context of the formal rules, the precise meaning of terms such as oxidation number and coordination number, and the conventions governing the use of multiplying affixes, enclosing marks, numbers and letters. The most important tables have been placed together at the end of the rules and numbered, and an index has been added.

0. PREAMBLE

Often the general principles of nomenclature do not stand out clearly in the detail of specific rules. The purpose of this preamble is to point out some general practices and to provide illustrative examples of the ways in which they are applied. Nomenclature is not static. While the illustrations are drawn largely from current practice, reference must occasionally be made to past and even to projected usage. Consequently, the illustrations in this preamble are not all officially accepted; the rules themselves provide recommended practice.

0.1. OXIDATION NUMBER

The concept of oxidation number is interwoven in the fabric of inorganic chemistry in many ways, including nomenclature. Oxidation number is an empirical concept; it is not synonymous with the number of bonds to an atom. The oxidation number of an element in any chemical entity is the charge which would be present on an atom of the element if the electrons in each bond to that atom were assigned to the more electronegative atom, thus:

		Oxidation Numbers	
MnO_4^-	= One Mn^{7+} and four O^{2-} ions	Mn = VII	O = −II
ClO^-	= One Cl^+ and one O^{2-} ion	Cl = I	O = −II
CH_4	= One C^{4-} and four H^+ ions	C = −IV	H = I
CCl_4	= One C^{4+} and four Cl^- ions	C = IV	Cl = −I
NH_4^+	= One N^{3-} and four H^+ ions	N = −III	H = I
NF_4^+	= One N^{5+} and four F^- ions	N = V	F = −I
AlH_4^-	= One Al^{3+} and four H^- ions	Al = III	H = −I
$[PtCl_2(NH_3)_2]$	= one Pt^{2+} and two Cl^- ions and two uncharged NH_3 molecules	Pt = II	Cl = −I
$[Ni(CO)_4]$	= one uncharged Ni atom and four uncharged CO molecules	Ni = 0	

By convention hydrogen is considered positive in combination with non-metals. For conventions concerning the oxidation numbers of organic radicals and the nitrosyl group see **7.313** and **7.323** respectively.

In the elementary state the atoms have oxidation state zero and a bond between atoms of the same element makes no contribution to the oxidation number, thus:

		Oxidation Numbers	
P_4	= four uncharged P atoms	P = 0	
P_2H_4	= two P^{2-} and four H^+ ions	P = −II	H = I
C_2H_2	= two C^- and two H^+ ions	C = −I	H = I
O_2F_2	= two O^+ and two F^- ions	O = I	F = −I
$Mn_2(CO)_{10}$	= two uncharged Mn atoms and ten uncharged CO molecules	Mn = 0	

Difficulties in assigning oxidation numbers may arise if the elements in a compound have similar electronegativities, *e.g.*, as in NCl_3 and S_4N_4.

0.2. COORDINATION NUMBER

The coordination number of the central atom in a compound is the number of atoms which are directly linked to the central atom. The attached atoms may be charged or uncharged or part of an ion or molecule. In some types of coordination compounds, the two atoms of a multiple bond in an attached group are assigned to a single coordination position. Crystallographers define the coordination number of an atom or ion in a lattice as the number of near neighbours to that atom or ion.

0.3. USE OF MULTIPLYING AFFIXES, ENCLOSING MARKS, NUMBERS AND ITALIC LETTERS

Chemical nomenclature uses multiplying affixes, numbers (both Arabic and Roman) and letters to indicate both stoicheiometry and structure.

0.31—Multiplying Affixes

The simple multiplying affixes, mono, di, tri, tetra, penta, hexa, hepta, octa, nona (ennea), deca, undeca (hendeca), dodeca, *etc.*, indicate

(a) stoicheiometric proportions

Examples:

CO	carbon monoxide
CO_2	carbon dioxide
P_4S_3	tetraphosphorus trisulfide

(b) extent of substitution

Examples:

$SiCl_2H_2$	dichlorosilane
$PO_2S_2^{3-}$	dithiophosphate ion

(c) number of identical coordinated groups

Example:

$[CoCl_2(NH_3)_4]^+$ tetraamminedichlorocobalt(III) ion

It is necessary in some languages to supplement these numerical affixes with hemi ($\frac{1}{2}$) and sesqui ($\frac{3}{2}$). The spelling eicosa (twenty) and icosa are both used.

These affixes have also somewhat different uses in designating

(1) the number of identical central atoms in condensed acids and their characteristic anions

Examples:

H_3PO_4	(mono)phosphoric acid
$H_4P_2O_7$	diphosphoric acid
$H_2S_3O_{10}$	trisulfuric acid

(2) the number of atoms of the same element forming the skeletons of some molecules or ions

Examples:

Si_2H_6	disilane
$B_{10}H_{14}$	decaborane(14)
$S_4O_6^{2-}$	tetrathionate ion

The multiplicative affixes bis, tris, tetrakis, pentakis, *etc.* were originally introduced into organic nomenclature to indicate a set of identical radicals each substituted in the same way

Examples:

$(ClCH_2CH_2)_2NH$	bis(2-chloroethyl)amine
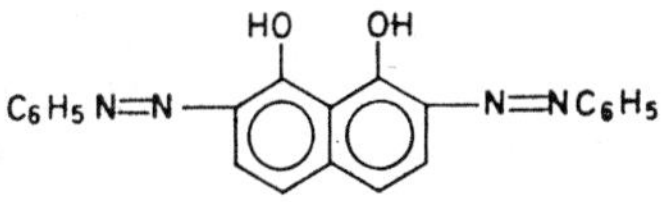 HO OH $C_6H_5N{=}N$— —$N{=}NC_6H_5$	2,7-bis(phenylazo)-1,8-naphthalenediol

or to avoid ambiguity

Examples:

$OC{=}CHC_6H_4CH{=}CO$	*p*-phenylenebisketene
$P(C_{10}H_{21})_3$	tris(decyl)phosphine

In the first case, bis is used to avoid confusion with the trivial name diketene used to describe the dimer of ketene. In the second case, tris(decyl) avoids any ambiguity with the organic radical tridecyl, $C_{13}H_{27}$. However, the use of these affixes has been extended, especially by *Chemical Abstracts*, to "all complex expressions".

Examples:

$Bi\{[(CH_3)_2NCH_2]NH\}_3$ tris{[(dimethylamino)methyl]amino}bismuthine
$[P(CH_2OH)_4]$ Cl tetrakis(hydroxymethyl)phosphonium chloride

Examples in inorganic nomenclature are:
$[Fe(CN)_2(CH_3NC)_4]$ dicyanotetrakis(methyl isocyanide)iron(II)
$Ca_5F(PO_4)_3$ pentacalcium fluoride tris(phosphate).

In the first inorganic example one wishes to avoid any doubt that the ligand is CH_3NC. In the latter, one must distinguish a double salt with phosphate from a salt of the condensed acid, triphosphate $[P_3O_{10}]^{5-}$.

Chemists are not agreed on the use of the multiplicative affixes bis, tris, *etc.* Some limit their use as far as possible and others follow the practice of "when in doubt use them." The former would restrict the use of bis, tris, *etc.*, to expressions containing another numerical affix, *e.g.*, bis(dimethylamino) and to cases where their absence would cause ambiguity, *e.g.*, trisdecyl for $(C_{10}H_{21})_3$ instead of tridecyl which is $(C_{13}H_{27})$. (*cf.* "Enclosing Marks" below).

In inorganic nomenclature, the affixes bi, ter, quater, *etc.* are used only in the names of a few organic molecules or radicals which may combine with inorganic compounds. In inorganic literature Latin affixes bi, tri (ter), quadri....multi are used in combination, *e.g.* with dentate (**7.1**) and valent, although the affixes di, tri, tetra....poly of Greek origin are used almost as frequently even with words of Latin origin.

0.32—Enclosing Marks

Enclosing marks are used in formulae to enclose sets of identical groups of atoms: $Ca_3(PO_4)_2$, $B[(CH_3)_2N]_3$. In names, enclosing marks generally are used following bis, tris, *etc.* around all complex expressions, and elsewhere to avoid any possibility of ambiguity (**7.21, 7.311, 7.314**). As with the case of multiplicative affixes, there are two attitudes toward the use of enclosing marks. Some use them only when absolutely necessary, while others use them every time a multiplicative affix is used. For example, the former omit enclosing marks in such cases as trisdecyl, tetrakishydroxymethyl, and tridecyl. The normal nesting order for enclosing marks is {[( )]}. However, in the formulae of coordination compounds, square brackets are used to enclose a complex ion or a neutral coordination entity. Enclosing marks are then nested with the square brackets as follows: [( )], [{( )}], [{[( )]}], [{{[( )]}}], *etc.*, and a space is left between the outer square brackets with their associated subscript or superscript symbol if any, and the remainder of the formula.

Examples:

$K_3\,[Co(C_2O_4)_3]$

$[Co(NO_2)_3(NH_3)_3]$

$[Co\{SC(NH_2)_2\}_4]\,[NO_3]_2$

$[Co\{SC[NH(CH_3)]_2\}_4]\,[NO_3]_2$

$[CoCl_2(NH_3)_4]_2\,SO_4$

0.33—Numbers

In names of inorganic compounds, Arabic numerals are used as locants to designate the atoms at which there is substitution, replacement, or addition in a chain, ring or cluster of atoms (*cf.* **7.72, 11.2, 11.4**). A standard pattern for assigning locants to each skeletal arrangement of atoms is necessary.

Examples:

$H_3Si—SiH_2—SiHCl—SiH_3$	2-chlorotetrasilane
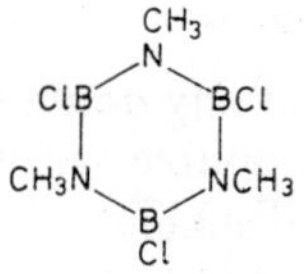	2,4,6-trichloro-1,3,5-trimethylborazine
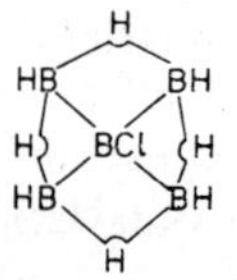	1-chloropentaborane(9)

Arabic numerals followed by + and − and enclosed in parentheses also are used to indicate the charge on a free or coordinated ion (**2.252**). Zero is not used with the name of an uncharged coordination compound. Finally Arabic numerals are sometimes used in place of numerical prefixes.

Examples:

$AlK(SO_4)_2 \cdot 12H_2O$	aluminium potassium sulfate 12-water
$8H_2S \cdot 46H_2O$	hydrogen sulfide-water(8/46)
B_6H_{10}	hexaborane(10)

Roman numerals are used in parentheses to indicate the oxidation number (or state) of an element (**2.252**). The cipher 0 is used to indicate an oxidation state of zero. A negative oxidation state is indicated by the use of negative sign with a Roman numeral (**2.252,** example 10; **7.323,** examples 5, 7).

0.34—Italic Letters

The symbols of the elements, printed in italics, are used to designate

(a) the element in a heteroatomic chain or ring at which there is substitution

CH_3ONH_2 *O*-methylhydroxylamine

```
   CH2CH2
  /      \
O          NCH3
  \      /
   CH2CH2
```

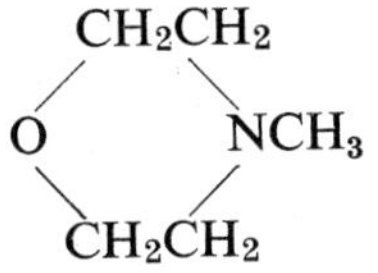

N-methylmorpholine

(b) the element in a ligand which is coordinated to a central atom (**7.33**)

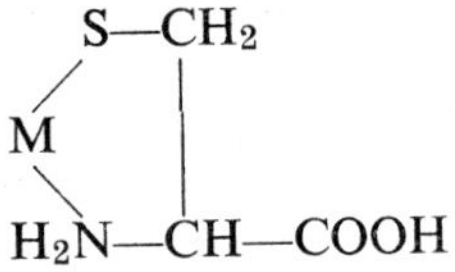

```
    S—CH2
   /    |
  M     |
   \    |
  H2N—CH—COOH
```

cysteinato-*S*,*N* . . . .

(c) the presence of bonds between two metal atoms (**7.712**)

$(OC)_3Fe(C_2H_5S)_2Fe(CO)_3$ bis(μ-ethylthio)bis(tricarbonyliron) (*Fe–Fe*).

(d) the point of attachment in some addition compounds

$CH_3ONH_2{\cdot}BH_3$ *O*-methylhydroxylamine (*N–B*)borane

(e) a specific isotope in isotopically labelled compounds (**1.32**)

$^{15}NH_3$ ammonia[^{15}N]

Lower case letters printed in italics are used as locants for the spatial positions around the central atom in a coordination compound (**7.512, 7.513, 7.514**). This usage is extended to multinuclear coordination compounds (**7.613**).

1. ELEMENTS

1.1. NAMES AND SYMBOLS OF THE ELEMENTS

1.11—The elements bear the symbols given in Table I*. It is desirable that the names differ as little as possible between different languages. The English list is given in Table I (p. 98). The isotopes ^{2}H and ^{3}H are named deuterium and tritium and the symbols D and T respectively may be used (**1.15**).

1.12—The names placed in parentheses (after the trivial names) in Table I shall always be used when forming names derived from those of the elements, *e.g.*, aurate, ferrate, wolframate and not goldate, ironate, tungstate.

For some compounds of sulfur, nitrogen and antimony derivatives of the Greek name *θεῖον*, the French name azote, and the Latin name stibium respectively, are used.

Although the name nickel is in accordance with the chemical symbol, it is essentially a trivial name, and is spelt so differently in various languages (niquel, nikkel, *etc.*) that it is recommended that derived names be formed from the Latin name niccolum, *e.g.*, niccolate in place of nickelate. The name mercury should be used as the root name also in languages where the element has another name (mercurate, *not* hydrargyrate).

In the cases in which different names have been used, the Commission has selected one based upon considerations of prevailing usage and practicability. It should be emphasized that their selection carries no implication regarding priority of discovery.

1.13—Any new metallic elements should be given names ending in -ium. Molybdenum and a few other elements have long been spelt without an "i" in most languages, and the Commission hesitates to insert it.

1.14—All new elements shall have 2-letter symbols.

1.15—All isotopes of an element except hydrogen should bear the same name. For hydrogen the isotope names protium, deuterium and tritium with the symbols ^{1}H, ^{2}H or D, and ^{3}H or T respectively are used. The prefixes deuterio- and tritio- are used when protium has been replaced. It is undesirable to assign isotopic names instead of numbers to other elements. They should be designated by mass numbers as, for example, "oxygen-18", with the symbol ^{18}O (**1.31**).

1.2. NAMES FOR GROUPS OF ELEMENTS, AND THEIR SUBDIVISIONS

1.21—The use of the collective names: halogens (F, Cl, Br, I and At), chalcogens (O, S, Se, Te, and Po), and halogenides (or halides) and chalcogenides for their compounds, alkali metals (Li to Fr), alkaline-earth metals

* If single letter symbols for the elements of iodine and vanadium are found inconvenient, *e.g.* in machine registration, the symbols Id and Va may be used.

(Ca to Ra), and noble gases may be continued. The use of the collective names "pnicogen" (N, P, As, Sb and Bi) and "pnictides" is not approved. The name rare-earth metals may be used for the elements Sc, Y, and La to Lu inclusive. The name lanthanoids for the elements 57–71 (La to Lu inclusive) is recommended; the names actinoids, uranoids, and curoids should be used analogously.

A transition element is an element whose atom has an incomplete *d* sub-shell, or which gives rise to a cation or cations with an incomplete *d* sub-shell.

When it is desired to designate sub-groups of the elements by the capital letters A and B these should be used as follows:

1A	2A	3A	4A	5A	6A	7A
K	Ca	Sc	Ti	V	Cr	Mn
Rb	Sr	Y	Zr	Nb	Mo	Tc
Cs	Ba	La*	Hf	Ta	W	Re
Fr	Ra	Ac†				

1B	2B	3B	4B	5B	6B	7B
Cu	Zn	Ga	Ge	As	Se	Br
Ag	Cd	In	Sn	Sb	Te	I
Au	Hg	Tl	Pb	Bi	Po	At

* Including the lanthanoids
† Including the actinoids, but thorium, protactinium and uranium may also be placed in groups 4A, 5A and 6A.

1.22—Because of the inconsistent uses in different languages of the word "metalloid" its use should be abandoned.

Elements should be classified as metals, semi-metals and non-metals.

1.3. INDICATION OF MASS, CHARGE, ETC., ON ATOMIC SYMBOLS

1.31—The mass number, atomic number, number of atoms, and ionic charge of an element may be indicated by means of four indices placed around the symbol. The positions are to be occupied thus:

left upper index .... mass number
left lower index .... atomic number
right lower index .... number of atoms
right upper index .... ionic charge

Ionic charge should be indicated by A^{n+} rather than by A^{+n}.

Example:

${}^{32}_{16}S_2^{2+}$ represents a doubly ionized molecule containing two atoms of sulfur, each of which has the atomic number 16 and mass number 32.

The following is an example of an equation for a nuclear reaction:

$${}^{26}_{12}Mg + {}^{4}_{2}He = {}^{29}_{13}Al + {}^{1}_{1}H$$

1.32—Isotopically labelled compounds may be described by inserting the symbol of the isotope in brackets into the name of the compound.

Examples:

1. $^{32}PCl_3$ phosphorus[^{32}P] trichloride (spoken: phosphorus-32 trichloride)
2. $H^{36}Cl$ hydrogen chloride[^{36}Cl] (spoken: hydrogen chloride-36)
3. $^{15}NH_3$ ammonia[^{15}N] (spoken: ammonia nitrogen-15)
4. $^{15}N^2H_3$ ammonia [^{15}N, ^2H] (spoken: ammonia nitrogen-15 hydrogen-2)
5. $^2H_2{}^{35}SO_4$ sulfuric[^{35}S] acid[^2H] (spoken: sulfuric acid sulfur-35 hydrogen-2)

If this method gives names which are ambiguous or difficult to pronounce, the whole group containing the labelled atom may be indicated.

Examples:

6. $HOSO_2{}^{35}SH$	thiosulfuric[^{35}SH] acid
7. $^{15}NO_2NH_2$	nitramide[$^{15}NO_2$], not nitr[^{15}N]amide
8. $NO_2{}^{15}NH_2$	nitramide[$^{15}NH_2$]
9. $HO_3S^{18}O{-}^{18}OSO_3H$	peroxo[$^{18}O_2$]disulfuric acid

1.4. ALLOTROPES

If systematic names for gaseous and liquid modifications are required, they should be based on the size of the molecule, which can be indicated by Greek numerical prefixes (listed in **2.251**). If the number of atoms be great and unknown, the prefix poly may be used. To indicate structures the prefixes in Table III may be used.

Examples:

	Symbol	*Trivial name*	*Systematic name*
1.	H	atomic hydrogen	monohydrogen
2.	O_2	(common) oxygen	dioxygen
3.	O_3	ozone	trioxygen
4.	P_4	white phosphorus (yellow phosphorus)	*tetrahedro*-tetraphosphorus
5.	S_8	λ-sulfur	*cyclo*-octasulfur or octasulfur
6.	S_n	μ-sulfur	*catena*-polysulfur or polysulfur

For the nomenclature of solid allotropic forms the rules in Section **10** may be applied.

2. FORMULAE AND NAMES OF COMPOUNDS IN GENERAL

Many chemical compounds are essentially binary in nature and can be regarded as combinations of ions or radicals; others may be treated as such for the purposes of nomenclature.

Some chemists have expressed the opinion that the name of a compound should indicate whether it is ionic or covalent. Such a distinction is made in some languages (*e.g.*, in German: Natriumchlorid *but* Chlorwasserstoff), but it has not been made in a consistent way, and indeed it seems impossible to introduce this distinction into a consistent system of nomenclature, because the line of demarcation between these two categories is not sharp. In these rules a system of nomenclature has been built upon the basis of the endings -ide and -ate, and it should be emphasized that these are intended to be applied both to ionic and covalent compounds. If it is desired to avoid such endings for neutral molecules, names can be given as coordination compounds in accordance with **2.24** and Section **7**.

2.1. FORMULAE

2.11—Formulae provide the simplest and clearest method of designating inorganic compounds. They are of particular importance in chemical equations and in descriptions of chemical procedure. Their general use in text matter is, however, not recommended, although in various circumstances a formula, on account of its compactness, may be preferable to a cumbersome and awkward name.

Formulae are often used to demonstrate structural connexions between atoms, or to provide other comparative chemical information. They may be written in any way necessary for such purposes. Nevertheless some standardization of formulae is desirable and the following rules codify the most widely accepted practices for writing formulae of inorganic compounds.

2.12—The *empirical formula* is formed by juxtaposition of the atomic symbols to give the *simplest possible* formula expressing the stoicheiometric composition of the compound in question. The empirical formula may be supplemented by indication of the crystal structure—see Section **10.**

2.13—For compounds consisting of discrete molecules the *molecular formula, i.e.*, a formula in accordance with the correct molecular weight of the compound, should be used, *e.g.*, S_2Cl_2 and $H_4P_2O_6$ and not SCl and H_2PO_3. When the molecular weight varies with temperature, *etc.*, the simplest possible formula may generally be chosen, *e.g.*, S, P, and NO_2 instead of S_8, P_4, and N_2O_4, unless it is desirable to indicate the molecular complexity.

2.14—In the *structural formula* the sequence and spatial arrangement of the atoms in a molecule is indicated.

2.15—In formulae the *electropositive constituent* (cation) should always be placed first*, *e.g.*, KCl, $CaSO_4$.

If the compound contains more than one electropositive or more than one electronegative constituent, the sequence within each class should be in alphabetical order of their symbols. Acids are treated as hydrogen salts, *e.g.*, H_2SO_4 and H_2PtCl_6; for the position of hydrogen see **6.323**, *cf.* **6.2**. To determine the position of complex ions only the symbol of the central atom is considered. For coordination compounds see **7.2**.

2.161—In the case of binary compounds between non-metals, in accordance with established practice, that constituent should be placed first which appears earlier in the sequence:

Rn, Xe, Kr, B, Si, C, Sb, As, P, N, H, Te, Se, S, At, I, Br, Cl, O, F.

Examples:

XeF_2, NH_3, H_2S, S_2Cl_2, Cl_2O, OF_2.

2.162—For chain compounds containing three or more elements, however, the sequence should in general be in accordance with the order in which the atoms are actually bound in the molecule or ion, *e.g.*, NCS^-, not CNS^-, HOCN (cyanic acid), and HONC (fulminic acid).

2.163—If two or more different atoms or groups are attached to a single central atom, the symbol of the central atom shall be placed first followed by the symbols of the remaining atoms or groups in alphabetical order, *e.g.*, $PBrCl_2$, $SbCl_2F$, PCl_3O, $P(NCO)_3O$, $PO(OCN)_3$. However, in the formulae of acids, hydrogen is placed first (see **5.2**). Deviations from this rule are also allowed when part of the molecule is regarded as a radical (see **3.3**).

2.17—In intermetallic compounds the constituents including Sb are placed in alphabetical order of their symbols. Deviations from this order may be allowed, for example to emphasize ionic character as in Na_3Bi_5, or when compounds with analogous structures are compared, *e.g.*, Cu_5Zn_8 and Cu_5Cd_8. In similar compounds containing non-metals, *e.g.*, interstitial compounds, metals are placed in alphabetical order of symbols, followed by the non-metals at the end in the order prescribed in **2.161**, *e.g.*, $MnTa_3N_4$.

2.18—The number of identical atoms or atomic groups in a formula is indicated by means of Arabic numerals, placed below and to the right of the symbol, or symbols in parentheses ( ) or square brackets [ ], to which they refer. Crystal water and similar loosely bound molecules are, however, enumerated by means of Arabic numerals before their formula.

Examples:

1. $CaCl_2$ not $CaCl^2$
2. $[Co(NH_3)_6]Cl_3$ not $[Co\ 6NH_3]\ Cl_3$
3. $[Co(NH_3)_6]_2\ [SO_4]_3$
4. $Na_2SO_4 \cdot 10H_2O$

2.19—The structural prefixes which may be used are listed in Table III. These should be connected with the formula by a hyphen and be italicized.

Example: *cis*-$[PtCl_2(NH_3)_2]$

* This also applies in Romance languages even though the electropositive constituent is placed last in the *name*, *e.g.*, KCl, chlorure de potassium.

2.2. SYSTEMATIC NAMES

Systematic names for compounds are formed by indicating the constituents and their proportions according to the following principles. Very many compounds consist, or can be regarded as essentially consisting, of two parts (binary type) as is expressed in the section on formulae **2.15–2.17**, and these can be dealt with by **2.21–2.24.**

2.21—The name of the *electropositive constituent* (or that treated as such according to **2.161**) will not be modified* (see, however, **2.2531**).

If the compound contains, or is regarded as containing, two or more electropositive constituents these should be cited in the order given in **6.31** and **6.32** (rules for double salts).

2.22—If the electronegative constituent is monoatomic or homopolyatomic its name is modified to end in -ide. For binary compounds the name of the element standing later in the sequence in **2.161** is modified to end in -ide. Elements other than those in the sequence of **2.161** are taken in the reverse sequence of Table IV and the name of the element occurring last is modified to end in -ide. If one element occurs only in Table IV and the other occurs also in the sequence of **2.161**, that from the sequence of **2.161** is modified to end in -ide.

Examples: Sodium plumbide, platinum bismuthide, potassium triiodide, sodium chloride, calcium sulfide, lithium nitride, arsenic selenide, calcium phosphides, nickel arsenide, aluminium borides, iron carbides, boron hydrides, phosphorus hydrides, hydrogen chloride, hydrogen sulfide, silicon carbide, carbon disulfide, sulfur hexafluoride, chlorine dioxide, oxygen difluoride.

Certain heteropolyatomic groups are also given the ending -ide, see **3.22.**

In Romance languages the endings -ure, -uro or -eto are used instead of -ide. In some languages the word *oxyde* is used, whereas the ending -ide is used in the names of other binary compounds; it is recommended that the ending -ide be universally adopted in these languages.

2.23—If the electronegative constituent is heteropolyatomic it should be designated by the termination -ate.

In certain exceptional cases the terminations -ide and -ite are used, for which see **3.22.**

In the case of two or more electronegative constituents their sequence of citation should be in alphabetical order (**6.33**).

2.24—In inorganic compounds it is generally possible in a polyatomic group to indicate a *characteristic atom* (as Cl in ClO^-) or a *central atom* (as I in ICl_4^-). Such a polyatomic group is designated as a *complex*, and the atoms, radicals, or molecules bound to the characteristic or central atom are termed *ligands*.

In this case the name of a negatively charged complex should be formed from the name of the characteristic or central element (as indicated in **1.12**) modified to end in -ate.

Anionic ligands are indicated by the termination -o. Further details

* In Germanic languages the electropositive constituent is placed first; but in Romance languages it is customary to place the electronegative constituent first.

concerning the designation of ligands, and the definition of "central atom", *etc.*, appear in Section **7.**

Although the terms sulfate, phosphate, *etc.*, were originally the names of the anions of particular oxoacids, the names sulfate, phosphate, *etc.*, should now designate quite generally a negative group containing sulfur or phosphorus respectively as the central atom, irrespective of its state of oxidation and the number and nature of the ligands. The complex is indicated by square brackets [ ], but this is not always necessary.

Examples:

1. Na_2 [SO_4]	disodium tetraoxosulfate
2. Na_2 [SO_3]	disodium trioxosulfate
3. Na_2 [S_2O_3]	disodium trioxothiosulfate
4. Na [SFO_3]	sodium fluorotrioxosulfate
5. Na_3 [PO_4]	trisodium tetraoxophosphate
6. Na_3 [PS_4]	trisodium tetrathiophosphate
7. Na [PCl_6]	sodium hexachlorophosphate
8. K [PF_2O_2]	potassium difluorodioxophosphate
9. K [PCl_2(NH)O]	potassium dichloroimidooxophosphate

In many cases these names may be abbreviated, *e.g.*, sodium sulfate, sodium thiosulfate (see **2.26**), and in other cases trivial names may be used (*cf.* **2.3**, **3.224**, and Section **5**). It should, however, be pointed out that the principle is quite generally applicable to compounds containing organic ligands also, and its use is recommended in all cases where trivial names do not exist.

Compounds of greater complexity to which the foregoing rules cannot be applied will be named according to Section 7. In many cases either system is possible, and when a choice exists the simpler name is preferable.

2.25—Indication of the Proportions of the Constituents

2.251—The stoicheiometric proportions may be denoted by means of Greek numerical prefixes (mono, di, tri, tetra, penta, hexa, hepta, octa, ennea, deca, hendeca, and dodeca) preceding without hyphen the names of the elements to which they refer. The Latin prefixes nona and undeca are also allowed. If the number of atoms be great and unknown, the prefix poly may be used. It may be necessary in some languages to supplement these numerals with hemi ($\frac{1}{2}$) and the Latin sesqui ($\frac{3}{2}$).

The prefix mono may be omitted except where confusion would arise. This is especially so in the case of certain ternary compounds where otherwise the extent of replacement of oxygen by some other element might be uncertain, *e.g.* CSO_2^{2-}, monothiocarbonate. Beyond 10, Greek prefixes may be replaced by Arabic numerals (with or without hyphen according to the custom of the language), as they are more readily understood. The end-vowels of numerical prefixes should not be elided except for compelling linguistic reasons.

The system is applicable to all types of compounds and is especially suitable for binary compounds of the non-metals.

When the number of complete groups of atoms requires designation, particularly when the name includes a numerical prefix with a different significance, the multiplicative numerals (Latin bis, Greek tris, tetrakis, *etc.*)

are used and the whole group to which they refer is placed in parentheses. (see Preamble).

Examples:

1. N_2O	dinitrogen oxide
2. NO_2	nitrogen dioxide
3. N_2O_4	dinitrogen tetraoxide
4. N_2S_5	dinitrogen pentasulfide
5. S_2Cl_2	disulfur dichloride
6. Fe_3O_4	triiron tetraoxide
7. U_3O_8	triuranium octaoxide
8. MnO_2	manganese dioxide
9. $Ca_3(PO_4)_2$	tricalcium bis(orthophosphate)
10. $Ca[PCl_6]_2$	calcium bis(hexachlorophosphate)

In indexes it may be convenient to italicize a numerical prefix at the beginning of the name and connect it to the rest of the name with a hyphen, *e.g.*, *tri*-Uranium octaoxide, but this is not desirable in a text.

Since the degree of polymerization of many substances varies with temperature, state of aggregation, *etc.*, the name to be used shall normally be based upon the simplest possible formula of the substance except when it is known to consist of small discrete molecules or when it is required specifically to draw attention to the degree of polymerization.

Example:

The name nitrogen dioxide may be used for the equilibrium mixture of NO_2 and N_2O_4. Dinitrogen tetraoxide means specifically N_2O_4.

2.252—The proportions of the constituents may also be indicated indirectly either by STOCK's system or by the EWENS-BASSETT system.

In STOCK's system the oxidation number of an element is indicated by a Roman numeral placed in parentheses immediately following the name of the element. For zero the cipher 0 is used. When used in conjunction with symbols the Roman numeral may be placed above and to the right.

The STOCK notation can also be applied to cations and anions. In employing it, use of the Latin names of the elements (or Latin roots) is considered advantageous.

Examples:

1. $FeCl_2$	iron(II) chloride or ferrum(II) chloride
2. $FeCl_3$	iron(III) chloride or ferrum(III) chloride
3. MnO_2	manganese(IV) oxide
4. BaO_2	barium(II) peroxide
5. P_2O_5	phosphorus(V) oxide or diphosphorus pentaoxide
6. As_2O_3	arsenic(III) oxide or diarsenic trioxide
7. $Pb_2^{II}Pb^{IV}O_4$	dilead(II) lead(IV) oxide or trilead tetraoxide
8. $K_4[Ni(CN)_4]$	potassium tetracyanoniccolate(0)
9. $K_4[Fe(CN)_6]$	potassium hexacyanoferrate(II)
10. $Na_2[Fe(CO)_4]$	sodium tetracarbonylferrate(−II)

In the EWENS-BASSETT system the charge of an ion indicated by an Arabic numeral followed by the sign of the charge is placed in parenthesis immediately following the name of the ion.

Examples:

11. $K_4[Ni(CN)_4]$ potassium tetracyanoniccolate(4−)
12. $K_4[Fe(CN)_6]$ potassium hexacyanoferrate(4−)
13. $Na_2[Fe(CO)_4]$ sodium tetracarbonylferrate(2−)
14. $Na_2N_2O_2$ sodium dioxodinitrate(2−)
15. $FeCl_2$ iron(2+) chloride
16. Hg_2Cl_2 dimercury(2+) chloride
17. UO_2SO_4 uranyl(2+) sulfate
18. $(UO_2)_2SO_4$ uranyl(1+) sulfate
19. $KReO_4$ potassium tetraoxorhenate(1−)

2.253—The following systems are in use but are discouraged:

2.2531—The system of valency indication by means of the terminations -ous and -ic applied to the cation may be retained for elements exhibiting not more than two valencies.

2.2532—"*Functional*" nomenclature (such as "nitric anhydride" for N_2O_5) is not recommended (Section **5**).

2.26—In systematic names it is not always necessary to indicate stoicheiometric proportions. In many instances it is permissible to omit the numbers of atoms, oxidation numbers, *etc.*, when they are not required in the particular circumstances. For instance, these indications are not generally necessary with elements of essentially constant valency.

Examples:

1. sodium sulfate instead of sodium tetraoxosulfate(VI)
2. aluminium sulfate instead of aluminium(III) sulfate

2.3. HYDRIDES

Binary hydrogen compounds may be named by the principles of **2.2**. Volatile hydrides, except those of Group VII and of oxygen and nitrogen, may be named by citing the root name of the element as indicated below followed by the suffix -ane. If the molecule contains more than one atom of that element, the number is indicated by the appropriate Greek numerical prefix (see **2.251**).

Recognized exceptions are water, ammonia, hydrazine, owing to long usage. Phosphine, arsine, stibine and bismuthine are also allowed. However, for all molecular hydrides containing more than one atom of the element, "-ane" names should be used. Caution must be exercised to avoid conflict with names of saturated six-membered heterocyclic rings, for example trioxane and diselenane.*

Examples:

1. B_2H_6 diborane
2. Si_3H_8 trisilane
3. GeH_4 germane
4. Sn_2H_6 distannane
5. PbH_4 plumbane
6. P_2H_4 diphosphane
7. As_2H_4 diarsane
8. H_2S_5 pentasulfane
9. H_2S_n polysulfane
10. H_2Se_2 diselane
11. H_2Te_2 ditellane

* See *I.U.P.A.C. Nomenclature of Organic Chemistry*, Butterworths, London, 1971, p. 53, **B-1.1.**

2.4. TRIVIAL NAMES

Certain well-established trivial names for oxoacids (Section **5**) and for hydrogen compounds (**2.3**) are still acceptable.

Purely trivial names, free from false scientific implications, such as soda, Chile saltpetre, quicklime are harmless in industrial and popular literature; but obsolete scientific names such as sulfate of magnesia, Natronhydrat, sodium muriate, carbonate of lime, *etc*., should be avoided under all circumstances, and they should be eliminated from technical and patent literature.

B

3. NAMES FOR IONS AND RADICALS

3.1. CATIONS

3.11—Monoatomic cations should be named as the corresponding element, without change or suffix, except as provided by **2.2531**.

Examples:

1. Cu^{+}	copper(I) ion
2. Cu^{2+}	copper(II) ion
3. I^{+}	iodine(I) cation

3.12—The preceding principle should also apply to polyatomic cations corresponding to radicals for which special names are given in **3.32**, *i.e.*, these names should be used without change or suffix.

Examples:

1. NO^{+}	nitrosyl cation
2. NO_2^{+}	nitryl cation

3.13—Polyatomic cations formed from monoatomic cations by the addition of other ions or neutral atoms or molecules (ligands) will be regarded as complex and will be named according to the rules given in Section **7**.

Examples:

1. $[Al(H_2O)_6]^{3+}$	hexaaquaaluminium ion
2. $[CoCl(NH_3)_5]^{2+}$	pentaamminechlorocobalt(2+) ion

For some important polyatomic cations which fall into this section, radical names given in **3.32** may be used alternatively, *e.g.*, for UO_2^{2+} the name uranyl(VI) ion in place of dioxouranium(VI) ion.

3.14—Names for polyatomic cations derived by addition of more protons than required to give a neutral unit to monoatomic anions, are formed by adding the ending -onium to the root of the name of the anion element (for nitrogen see **3.15**).

Examples:

phosphonium, arsonium, stibonium, oxonium, sulfonium, selenonium, telluronium, fluoronium and iodonium ions.

Ions derived by substitution in these parent cations should be named as such, whether the parent itself be a known compound or not. For example, PCl_4^{+}, the tetrachlorophosphonium ion, and $(CH_3)_4Sb^{+}$, the tetramethylstibonium ion.

The ion H_3O^{+}, which is in fact the monohydrated proton, is to be known as the oxonium ion when it is believed to have this constitution, as for example in $H_3O^{+}ClO_4^{-}$, oxonium perchlorate. If the hydration is of no particular importance to the matter under consideration, the simpler term hydrogen ion may be used. The latter may also be used for the indefinitely solvated proton in non-aqueous solvents; but definite ions such as $CH_3OH_2^+$ and $(CH_3)_2OH^+$ should be named as derivatives of the oxonium ion, *i.e.*, as methyl and dimethyl oxonium ions respectively.

3.15—Nitrogen Cations

3.151—The name ammonium for the ion NH_4^+ does not conform to **3.14**, but is retained. Substituted ammonium ions are named similarly, for example NF_4^+, the tetrafluoroammonium ion. This decision does *not* release the word nitronium for other uses: this would lead to inconsistencies when the rules were applied to other elements.

3.152—Substituted ammonium ions derived from nitrogen bases with names ending in -amine will receive names formed by changing -amine into -ammonium. For example, $HONH_3^+$ the hydroxylammonium ion.

3.153—When the nitrogen base is known by a name ending otherwise than in -amine, the cation name is to be formed by adding the ending -ium to the name of the base (if necessary omitting a final -e or other vowel).

Examples:

hydrazinium, anilinium, glycinium, pyridinium, guanidinium, imidazolium, *etc.*

The names uronium and thiouronium, derived from the names urea and thiourea respectively, though inconsistent with this rule, may be retained.

3.16—Cations formed by adding protons to non-nitrogenous bases may also be given names formed by adding -ium to the name of the compound to which the proton is added.

Examples:

dioxanium, acetonium.

In the case of cations formed by adding protons to acids, however, their names are to be formed by adding the word acidium to the name of the corresponding anion, and not from that of the acid itself. For example, $H_2NO_3^+$ the nitrate acidium ion; $H_2NO_2^+$ the nitrite acidium ion and $CH_3COOH_2^+$ the acetate acidium ion. Note, however, that when the anion of the acid is monatomic **3.14** will apply; for example, FH_2^+ is the fluoronium ion.

3.17—Where more than one ion is derived from one base, as, for example, $N_2H_5^+$ and $N_2H_6^{2+}$, their charges may be indicated in their names as the hydrazinium(1+) and the hydrazinium(2+) ion, respectively.

3.2. ANIONS

3.21—The names for monoatomic anions shall consist of the name (sometimes abbreviated) of the element, with the termination -ide. Thus:

H^-	hydride ion	O^{2-}	oxide ion	N^{3-}	nitride ion
D^-	deuteride ion	S^{2-}	sulfide ion	P^{3-}	phosphide ion
F^-	fluoride ion	Se^{2-}	selenide ion	As^{3-}	arsenide ion
Cl^-	chloride ion	Te^{2-}	telluride ion	Sb^{3-}	antimonide ion
Br^-	bromide ion			C^{4-}	carbide ion
I^-	iodide ion			Si^{4-}	silicide ion
				B^{3-}	boride ion

Expressions of the type of "chlorine ion" are used particularly in connection with crystal structure work and spectroscopy; the Commission recommends that whenever the charge corresponds to that indicated above, the termination -ide should be used.

3.22—Polyatomic Anions

3.221—Certain polyatomic anions have names ending in -ide. These are:

HO^-	hydroxide ion	N_3^-	azide ion
O_2^{2-}	peroxide ion	NH^{2-}	imide ion
O_2^-	hyperoxide ion	NH_2^-	amide ion
O_3^-	ozonide ion	$NHOH^-$	hydroxylamide ion
S_2^{2-}	disulfide ion	$N_2H_3^-$	hydrazide ion
I_3^-	triiodide ion	CN^-	cyanide ion
HF_2^-	hydrogendifluoride ion	C_2^{2-}	acetylide ion

Names for other polysulfide, polyhalogenide and like ions containing an element of one kind only may be formed in analogous manner. The HO^- ion should not be called the hydroxyl ion. The name hydroxyl is reserved for the HO group when neutral or positively charged, whether free or as a substituent (*cf.* **3.12** and **3.32**).

3.222—Ions such as HS^- and HO_2^- will be called the hydrogensulfide ion and the hydrogenperoxide ion respectively. This agrees with **6.2**, and names such as hydrosulfide are not required.

3.223—The names for other polyatomic anions shall consist of the root of the name of the central atom with the termination -ate **(1.12)**, which is used quite generally for complex anions. Atoms and groups attached to the central atom shall generally be treated as ligands in a complex (*cf.* **2.24** and Section **7**) as, for example, $[Sb(OH)_6]^-$ the hexahydroxoantimonate(v) ion.

This applies also when the exact composition of the anion is not known; *e.g.*, by solution of aluminium hydroxide or zinc hydroxide in sodium hydroxide, aluminate and zincate ions are formed.

3.224—It is quite practicable to treat oxygen also in the same manner as other ligands **(2.24)**, but it has long been customary to ignore the name of this element altogether in anions and to indicate its presence and proportion by means of a series of prefixes (hypo-, per-, *etc.*, see Section **5**) and sometimes also by the suffix -ite in place of -ate.

The termination -ite has been used to denote a lower state of oxidation and may be retained in trivial names in the following cases:

NO_2^-	nitrite	SO_3^{2-}	sulfite	ClO_2^-	chlorite
$N_2O_2^{2-}$	hyponitrite	$S_2O_5^{2-}$	disulfite	ClO^-	hypochlorite
				BrO^-	hypobromite
NOO_2^-	peroxonitrite	$S_2O_4^{2-}$	dithionite	IO^-	hypoiodite
		$S_2O_2^{2-}$	thiosulfite		
AsO_3^{3-}	arsenite	SeO_3^{2-}	selenite		

For the names of the ions formerly called phosphite, diphosphite and hypophosphite respectively, see Table II. Esters of the hypothetical acid $P(OH)_3$ are, however, called phosphites.

The Commission does not recommend the use of any such names other than those listed. A number of other names ending in -ite have been used, *e.g.*, antimonite, tellurite, stannite, plumbite, ferrite, manganite, *etc.*, but in many cases such compounds are known to be double oxides in the solid state and are to be treated as such, *cf.* **6.5**, *e.g.*, Cr_2CuO_4 chromium(III) copper(II) oxide, not copper chromite. Where there is a reason to believe that they denote a definite salt with a discrete anion the name is formed in accordance with **3.223**. By dissolving, for example, Sb_2O_3, SnO, or PbO in sodium hydroxide an antimonate(III), a stannate(II), a plumbate(II), *etc.*, is formed in the solution.

Concerning the use of prefixes hypo-, per-, *etc.*, see the list of acids, table in **5.214**. For all new compounds and even for the less common ones listed in the table in **3.224** or derived from the acids listed in the table in **5.214** it is preferable to use the system given in **2.24** and in Sections **5** and **7**.

3.3. RADICALS

3.31—A radical is here regarded as a group of atoms which occurs repeatedly in a number of different compounds. Sometimes the same radical fulfils different functions in different cases, and accordingly different names have often been assigned to the same group. The Commission considers it desirable to reduce this diversity and recommends that formulae or systematic names are applied to denote all new radicals, instead of introducing new trivial names. Table II gives an extensive selection of radical names at present in use in inorganic chemistry.

3.32—Certain neutral and cationic radicals (for anions see **3.22**) containing oxygen or other chalcogens have, irrespective of charge, special names ending in -yl, and the Commission approves the provisional retention of the following:

HO	hydroxyl	SO	sulfinyl (thionyl)	ClO	chlorosyl
CO	carbonyl	SO_2	sulfonyl (sulfuryl)	ClO_2	chloryl
NO	nitrosyl	S_2O_5	disulfuryl	ClO_3	perchloryl (and similarly for other halogens)
NO_2	nitryl*	SeO	seleninyl		
PO	phosphoryl	SeO_2	selenonyl		
		CrO_2	chromyl		
		UO_2	uranyl		
		NpO_2	neptunyl		
		PuO_2	plutonyl (similarly for other actinoid elements)		

* The name nitroxyl should not be used for this group since the name nitroxylic acid has been used for H_2NO_2. Although the word nitryl is firmly established in English, nitroyl may be a better model for many other languages.

Such names as the above should only be used to designate compounds consisting of discrete molecules. Names such as bismuthyl and antimonyl are not approved because the compounds do not contain BiO and SbO groups; such compounds are to be designated as oxides (**6.4**).

Radicals analogous to the above containing other chalcogens in place of oxygen are named by adding the prefixes thio-, seleno-, *etc.*

Examples:

PS thiophosphoryl CSe selenocarbonyl, *etc.*

In cases where the characteristic element of a radical may have different oxidation numbers these should be indicated by the STOCK notation; when the radical is an ion its charge may be indicated by the EWENS-BASSETT system. For example, the ions UO_2^{2+} and UO_2^{+} can be named as uranyl(VI) and uranyl(V) or as uranyl(2+) and uranyl(1+) respectively.

These polyatomic radicals are always treated as forming the positive part of the compound.

Examples:

1. $COCl_2$	carbonyl chloride	6. S_2O_5ClF	disulfuryl chloride fluoride
2. NOS	nitrosyl sulfide	7. $SO_2(N_3)_2$	sulfonyl (sulfuryl) diazide
3. PON	phosphoryl nitride	8. SO_2NH	sulfonyl (sulfuryl) imide
4. $PSCl_3$	thiophosphoryl chloride	9. IO_2F	iodyl fluoride
5. $NO_2HS_2O_7$	nitryl hydrogendisulfate		

By using the same radical names regardless of unknown or controversial polarity relationships stable names can be formed without entering upon controversy. Thus, for example, the compounds NOCl and $NOClO_4$ are quite unambiguously denoted by the names nitrosyl chloride and nitrosyl perchlorate respectively.

3.33—It should be noted that the same radical may have different names in inorganic and organic chemistry. To draw attention to such differences the prefix names of radicals as substituents in organic compounds have been listed together with the inorganic names in the list of names in Table II. Names of purely organic compounds, of which many are important in the chemistry of coordination compounds (Section **7**), should be in accordance with the nomenclature of organic chemistry.

Organic chemical nomenclature is to a large extent built upon the scheme of substitution, *i.e.*, replacement of hydrogen atoms by other atoms or groups. Such "substitutive names" are extremely rare in inorganic chemistry; they are used, *e.g.*, in the following cases: NH_2Cl is called chloramine, and $NHCl_2$ dichloramine. These names may be retained in absence of better terms. Other substitutive names (derived from "sulfonic acid" as a name for HSO_3H) are fluorosulfonic and chlorosulfonic acid, aminosulfonic acid, iminodisulfonic acid, and nitrilotrisulfonic acid. These names should be replaced by the following:

HSO_3F	fluorosulfuric acid	NH_2SO_3H	amidosulfuric acid
HSO_3Cl	chlorosulfuric acid	$NH(SO_3H)_2$	imidobis(sulfuric) acid
		$N(SO_3H)_3$	nitridotris(sulfuric) acid

Names such as chlorosulfuric acid and amidosulfuric acid might be considered as substitutive names, derived by substitution of *hydroxyl* groups in sulfuric acid. From a more fundamental point of view, however (see **2.24**), such names are formed by adding hydroxyl, amide, imide, *etc.*, groups together with oxygen atoms to a sulfur atom, "sulfuric acid" in this connection standing as an abbreviation for "trioxosulfuric acid".

Another type of organic nomenclature,* the formation of "conjunctive names", is also met only in a few cases in inorganic chemistry, *e.g.*, the hydrazinesulfonic and hydroxylaminesulfonic acids. According to the principles of inorganic chemical nomenclature these compounds should be called hydrazidosulfuric and hydroxylamidosulfuric acids.

* *cf. I.U.P.A.C. Nomenclature of Organic Chemistry*, Butterworths, London, 1971, p. 118, Rule **C–0.5.**

4. ISO- AND HETERO-POLYANIONS

4.1. ISOPOLYANIONS

4.11—Without recourse to structural information, salts containing polyanions may be given their complete stoicheiometric name, according to **2.24**.

Examples:

1. $Na_2S_2O_7$	disodium heptaoxodisulfate
2. $Na_2S_2O_5$	disodium pentaoxodisulfate
3. $K_2H_2P_2O_6$	dipotassium dihydrogenhexaoxodiphosphate
4. $Na_2Mo_6O_{18}$	disodium 18-oxohexamolybdate

4.12—Anions of polyacids derived by condensation of molecules of the same monoacid, containing the characteristic element in the oxidation state corresponding to its Group number, are named by indicating with Greek numerical prefixes the number of atoms of that element. It is not necessary to give the number of oxygen atoms when the charge of the anion or the number of cations is indicated.

Examples:

1. $S_2O_7^{2-}$	disulfate(2–)
2. $Si_2O_7^{6-}$	disilicate(6–)
3. $Te_4O_{14}^{4-}$	tetratellurate(4–)
4. $Cr_4O_{13}^{2-}$	tetrachromate(2–)
5. $P_3O_{10}^{5-}$	triphosphate(5–)
6. $Mo_7O_{24}^{6-}$	heptamolybdate(6–)
7. $Na_2B_4O_7$	disodium tetraborate
8. NaB_5O_8	sodium pentaborate
9. $Ca_3Mo_7O_{24}$	tricalcium heptamolybdate
10. $Na_7HNb_6O_{19} \cdot 15H_2O$	heptasodium monohydrogen hexaniobate-15-water
11. $K_2Mg_2V_{10}O_{28} \cdot 16H_2O$	dimagnesium dipotassium decavanadate-16-water

4.13—When the characteristic element is partially or wholly present in a lower oxidation state than corresponds to its Group number, its oxidation state(s) may be indicated by STOCK number(s). If evidence is available, more than one STOCK number may be used, and the lowest should be cited first.

Examples:

1. $[S_2O_5]^{2-}$	disulfate(IV)(2–) (trivial name disulfite)
2. $[O_2HP—O—PHO_2]^{2-}$	dihydrogendiphosphate(III)(2–) (trivial name diphosphonate)
3. $[O_2HP—O—PO_3H]^{2-}$	dihydrogendiphosphate(III, V)(2–)
4. $[HO_3P—PO_3H]^{2-}$	dihydrogendiphosphate(IV)(2–) (trivial name dihydrogenhypophosphate)
5. $[Mo_2{}^{V}Mo_4{}^{VI}O_{18}]^{2-}$	hexamolybdate(2V, 4VI)(2–)

4.14—Cyclic and straight chain structures may be distinguished by means of the prefixes *cyclo* and *catena*, although the latter may usually be omitted.*

Examples:

1. $\left[\begin{array}{c} \mathrm{O}\quad\ \mathrm{O}\quad\ \mathrm{O} \\ \mathrm{OP{-}O{-}P{-}O{-}PO} \\ \mathrm{O}\quad\ \mathrm{O}\quad\ \mathrm{O} \end{array}\right]^{5-}$ triphosphate

2. $\left[\begin{array}{ccc} & \mathrm{O_2P} & \\ \mathrm{O} & & \mathrm{O} \\ \mathrm{O_2P} & & \mathrm{PO_2} \\ & \mathrm{O} & \end{array}\right]^{3-}$ *cyclo*-triphosphate

3. $[\mathrm{O(PO_3)}_n]^{(n+2)-}$ *catena*-polyphosphate

4. $\left[\begin{array}{c} \mathrm{O}\quad\ \mathrm{O}\quad\ \mathrm{O} \\ \mathrm{OSi{-}O{-}Si{-}O{-}SiO} \\ \mathrm{O}\quad\ \mathrm{O}\quad\ \mathrm{O} \end{array}\right]^{8-}$ trisilicate

5. $\left[\begin{array}{ccc} & \mathrm{O_2Si} & \\ \mathrm{O} & & \mathrm{O} \\ \mathrm{O_2Si} & & \mathrm{SiO_2} \\ & \mathrm{O} & \end{array}\right]^{6-}$ *cyclo*-trisilicate

6. $\mathrm{K_4H_4Si_4O_{12}}$ tetrapotassium tetrahydrogen-*cyclo*-tetrasilicate

4.15—Anions corresponding to thioacids, peroxoacids, amidoacids, *etc.*, derived from isopolyanions are named by adding the prefixes thio, peroxo, and amido, *etc.*, to the name of the parent ion. If there is a possibility of isomerism, and the structure of the compound is known, the atom or atoms of the characteristic element, to which the group substituting oxygen is bonded, is indicated by numbers. For this purpose the atoms of the characteristic elements are numbered consecutively along the chain. The direction of numbering is chosen to give substituting atoms lowest possible locants;† substituents are cited in alphabetical order. Bridging atoms or groups are designated by μ, preceded if necessary by the locants of the atoms which are bridged (**7.61**).

Examples:

1. $\left[\begin{array}{c} \mathrm{O}\quad\ \mathrm{O}\quad\ \mathrm{O} \\ \mathrm{SP{-}O{-}P{-}O{-}PO} \\ \mathrm{O}\quad\ \mathrm{O}\quad\ \mathrm{O} \end{array}\right]^{5-}$ 1-thiotriphosphate(5–)

* In mineralogy and geochemistry silicates containing chains (single or double), sheets or three-dimensional frameworks are designated by the prefixes *ino*, *phyllo* or *tecto*, respectively.

Examples:

$(\mathrm{BaSiO_3})_x$ barium *ino*-polymetasilicate
$(\mathrm{Ca_3Si_4O_{11}})_x$ tricalcium *ino*-polytetrasilicate
$(\mathrm{Na_2Si_2O_5})_x$ disodium *phyllo*-polydisilicate
$[\mathrm{Mg_3(OH)_2Si_4O_{10}}]_x$ trimagnesium dihydroxide *phyllo*-polytetrasilicate

† *cf. I.U.P.A.C. Nomenclature of Organic Chemistry*, Butterworths, London, 1971, pp. 10–11, Rule A-**2.6.**

2. $\left[\begin{array}{c} \mathrm{O\quad\ \ O\qquad\ O} \\ \mathrm{OP{-}S{-}P{-}O{-}PO} \\ \mathrm{O\quad\ \ O\qquad\ O} \end{array}\right]^{5-}$ 1,2-*μ*-thiotriphosphate(5–)

3. $\left[\begin{array}{c} \mathrm{O\qquad\ S\qquad\ O} \\ \mathrm{OP{-}O{-}P{-}O{-}PO} \\ \mathrm{O\qquad\ O\qquad\ O} \end{array}\right]^{5-}$ 2-thiotriphosphate(5–)

4. $\left[\begin{array}{c} \mathrm{S\qquad\ S\qquad\ O} \\ \mathrm{OP{-}O{-}P{-}O{-}PO} \\ \mathrm{O\qquad\ O\qquad\ O} \end{array}\right]^{5-}$ 1,2-dithiotriphosphate(5–)

5. $\left[\begin{array}{c} \mathrm{S\qquad\ O\qquad\ O} \\ \mathrm{SP{-}O{-}P{-}O{-}PNH_2} \\ \mathrm{S\qquad\ O\qquad\ O} \end{array}\right]^{4-}$ 3-amido-1,1,1-trithiotriphosphate(4–)

6. $\left[\begin{array}{c} \mathrm{O\qquad\ O\qquad\ O} \\ \mathrm{H_2NP{-}O{-}P{-}O{-}PO} \\ \mathrm{O\qquad\ O\qquad\ O} \end{array}\right]^{4-}$ 1-amidotriphosphate(4–)

7. $\left[\begin{array}{c} \mathrm{O\qquad\qquad O\qquad\ O\qquad\ O} \\ \mathrm{OP{-}NH{-}P{-}O{-}P{-}O{-}PO} \\ \mathrm{O\qquad\qquad O\qquad\ O\qquad\ O} \end{array}\right]^{6-}$ 1,2-*μ*-imidotetraphosphate(6–)

8. $\left[\begin{array}{ccc} & \mathrm{O_2} & \\ & \mathrm{P} & \\ \mathrm{O} & & \mathrm{NH} \\ | & & | \\ \mathrm{O_2P} & & \mathrm{PO_2} \\ & \mathrm{O} & \end{array}\right]^{3-}$ *μ*-imido-*cyclo*-triphosphate(3–)

4.16—If most of the oxygen atoms of an oxoacid have been substituted, the name shall be formed according to **2.24** and **7.6**.

Examples:

1. $[F_5As{-}O{-}AsF_5]^{2-}$ decafluoro-*μ*-oxo-diarsenate(2–)
2. $[(O_2)_2OCr{-}O{-}O{-}CrO(O_2)_2]^{2-}$ 1,2-dioxo-*μ*-peroxo-1,1,2,2-tetraperoxodichromate(2–)
3. $[S_3P{-}O{-}PS_2{-}O{-}PS_3]^{5-}$ di-*μ*-oxo-octathiotriphosphate(5–)

4.2. HETEROPOLYANIONS

4.21—Heteropolyanions with a Chain or Ring Structure

4.211—Dinuclear anions are named by treating the anion which comes first in alphabetical order as the ligand on the characteristic atom of the second.

Examples:

1. $[O_3P{-}O{-}SO_3]^{3-}$ phosphatosulfate(3–)
2. $[O_3S{-}O{-}CrO_3]^{2-}$ chromatosulfate(2–)
3. $[O_3Se{-}O{-}SO_3]^{2-}$ selenatosulfate(2–)
4. $[O_3Cr{-}O{-}SeO_3]^{2-}$ chromatoselenate(2–)
5. $[O_3As{-}O{-}PO_3]^{4-}$ arsenatophosphate(4–)

4.212—Longer chains are named similarly (unless the chain contains an obvious central atom, when **4.214** applies), beginning with the end group which comes first in alphabetical order and treating the chain with (*n*—1) units as the ligand on the other end group.

Examples:

1. $[O_3Cr—O—AsO_2—O—PO_3]^{4-}$ (chromatoarsenato)phosphate(4–)
2. $[O_3Cr—O—PO_2—O—AsO_3]^{4-}$ (arsenatophosphato)chromate(4–)
3. $[O_3As—O—AsO_2—O—PO_3]^{5-}$ (diarsenato)phosphate(5–)
4. $[O_3S—O—CrO_2—O—AsO_2—O—PO_3]^{4-}$ [(phosphatoarsenato)chromato]-sulfate(4–)

4.213—Cyclic heteropolyanions are named in a manner similar to those with a chain structure, the starting point and direction of citation of the units being chosen according to alphabetical priority.

Examples:

```
1.  [       O2            ]2-
    [    /As—O\           ]
    [   O       PO2       ]
    [   |        |        ]
    [ O2Cr       O        ]
    [    \O — S/          ]
    [         O2          ]
```
cyclo-arsenatochromatosulfato-phosphate(2–)

```
2.  [        O2             ]2-
    [ O2Cr—O—As—O—CrO2      ]
    [   |           |       ]
    [   |           |       ]
    [   O—P——O——S—O         ]
    [     O2    O2          ]
```
cyclo-arsenatochromatophosphato-sulfatochromate(2–)

```
3.  [    /AsO2—O\         ]4-
    [   O        PO2      ]
    [   |         |       ]
    [ O2As        O       ]
    [    \O————As/        ]
    [          O2         ]
```
cyclo-triarsenatophosphate(4–)

4.214—In polyanions with an obvious central atom the peripheral anions are named as ligands on the central atom and cited in alphabetical order.

Examples:

```
1.  [      O       ]4-
    [ O3CrOPOAsO3  ]
    [      O       ]
    [     SO3      ]
```
(arsenato)(chromato)(sulfato)phosphate(4–)

2. $[OAs(MoO_4)_3]^{3-}$ tris(molybdato)arsenate(3–)
3. $[O_3As—O—PO_2—O—AsO_3]^{5-}$ bis(arsenato)phosphate(5–)
4. $[O_3P—O—AsO_2—O—PO_3]^{5-}$ bis(phosphato)arsenate(5–)

When the central atom has no oxo ligand this name is identical with that obtained by applying the nomenclature of coordination chemistry, *e.g.* $[B(ONO_2)_4]^-$, tetranitratoborate.

4.22—Condensed Heteropolyanions

The three-dimensional framework of linked WO_6, MoO_6, *etc.*, octahedra surrounding the central atom are designated by the prefixes wolframo, molybdo, *etc.*, *e.g.*, wolframophosphate (tungstophosphate), *not* phosphowolframate (phosphotungstate). The numbers of atoms of the characteristic element are indicated by Greek prefixes or numerals.

If the oxidation number has to be given, it may be necessary to place it immediately after the atom referred to and not after the ending -ate, in order to avoid ambiguity.

Examples:

1. $[PW_{12}O_{40}]^{3-}$	dodecawolframophosphate(3–) or 12-wolframophosphate(3–)
2. $[PMo_{10}V_2O_{39}]^{3-}$	decamolybdodivanadophosphate(3–)
3. $[Co^{II}Co^{III}W_{12}O_{42}]^{7-}$	dodecawolframocobalt(II)cobalt(III)ate
4. $[Mn^{IV}Mo_9O_{32}]^{6-}$	nonamolybdomanganate(6–)
5. $[Ni(OH)_6W_6O_{18}]^{4-}$	hexahydroxohexawolframoniccolate(4–)
6. $[IW_6O_{24}]^{5-}$	hexawolframoperiodate(5–)
7. $[Ce^{IV}Mo_{12}O_{42}]^{8-}$	dodecamolybdocerate(IV)(8–)
8. $[Cr^{III}Mo_6O_{21}]^{3-}$	hexamolybdochromate(III)(3–)
9. $[P^{V}_2Mo_{18}O_{62}]^{6-}$	18-molybdodiphosphate(V)(6–)
10. $[P^{III}_2Mo_{12}O_{41}]^{4-}$	dodecamolybdodiphosphate(III)(4–)
11. $[S^{IV}_2Mo_5O_{21}]^{4-}$	pentamolybdodisulfate(IV)(4–)

The names of salts and free acids are given in the usual way, *e.g.*:

12. $[NH_4]_6\,[TeMo_6O_{24}] \cdot 7H_2O$	hexaammonium hexamolybdotellurate heptahydrate
13. $Li_3H\,[SiW_{12}O_{40}] \cdot 24H_2O$	trilithium hydrogen dodecawolframosilicate-24-water
14. $H_4\,[SiW_{12}O_{40}]$	tetrahydrogen dodecawolframosilicate or dodecawolframosilicic acid

5. ACIDS

Many of the compounds which now according to some definitions are called acids do not fall into the classical province of acids. In other parts of inorganic chemistry functional names are disappearing and it would have been most satisfactory to abolish them also for those compounds generally called acids. Names for these acids may be derived from the names of the anions as in Section **2**, *e.g.*, hydrogen sulfate instead of sulfuric acid. The nomenclature of acids has, however, a long history of established custom, and it appears impossible to systematize acid names without drastic alteration of the accepted names of many important and well-known substances.

The present rules are aimed at preserving the more useful of the older names while attempting to guide further development along directions which should allow new compounds to be named in a more rational manner.

5.1. BINARY AND PSEUDOBINARY ACIDS

Acids giving rise to the -ide anions defined by **3.21** and **3.221** will be named as binary and pseudobinary compounds of hydrogen, *e.g.*, hydrogen chloride, hydrogen sulfide, hydrogen cyanide.

For the compound HN_3 the name hydrogen azide is recommended in preference to hydrazoic acid.

5.2. ACIDS DERIVED FROM POLYATOMIC ANIONS

Acids giving rise to anions bearing names ending in -ate or in -ite may also be treated as in **5.1**, but names more in accordance with custom are formed by using the term -ic acid and -ous acid corresponding to the anion terminations -ate and -ite respectively. Thus chloric acid corresponds to chlorate, sulfuric acid to sulfate, and nitrous acid to nitrite.

This nomenclature may also be used for less common acids, *e.g.*, hexacyanoferrate ions correspond to hexacyanoferric acids. In such cases, however, systematic names of the type hydrogen hexacyanoferrate are preferable.

Most of the common acids are oxoacids, *i.e.*, they contain only oxygen atoms bound to the characteristic atom. It is a long-established custom not to indicate these oxygen atoms. It is mainly for these acids that long-established names will have to be retained. Most anions may be considered as complexes and the acids given names in accordance therewith, for example H_4XeO_6, hexaoxoxenonic(VIII) acid or hydrogen hexaoxoxenonate(VIII).

5.21—Oxoacids

For the oxoacids the ous-ic notation to distinguish between different oxidation states is applied in many cases. The -ous acid names are restricted to acids corresponding to the -ite anions listed in the table in **3.224**.

Further distinction between different acids with the same characteristic element is in some cases effected by means of prefixes. This notation should not be extended beyond the cases listed below.

5.211—The prefix hypo- is used to denote a lower oxidation state, and may be retained in the following cases:

$HClO$	hypochlorous acid	$H_2N_2O_2$	hyponitrous acid
$HBrO$	hypobromous acid	$H_4P_2O_6$	hypophosphoric acid
HIO	hypoiodous acid		

5.212—The prefix per- has been used to designate a higher oxidation state and is retained only for $HClO_4$, perchloric acid, and corresponding acids of the other elements in Group VII. This use of the prefix per- should not be extended to elements of other Groups, and such names as perxenonate and perruthenate are not approved. The prefix per- should not be confused with the prefix peroxo-, see **5.22.**

5.213—The prefixes ortho- and meta- have been used to distinguish acids differing in the "content of water". The following names are approved:

H_3BO_3	orthoboric acid	$(HBO_2)_n$	metaboric acid
H_4SiO_4	orthosilicic acid	$(H_2SiO_3)_n$	metasilicic acid
H_3PO_4	orthophosphoric acid	$(HPO_3)_n$	metaphosphoric acid
H_5IO_6	orthoperiodic acid		
H_6TeO_6	orthotelluric acid		

The prefix pyro- has been used to designate an acid formed from two molecules of an ortho-acid minus one molecule of water. Such acids can now generally be regarded as the simplest cases of isopolyacids (*cf.* **4.12**). The trivial name pyrophosphoric acid may be retained for $H_4P_2O_7$, although diphosphoric acid is preferable.

5.214—The following table contains the accepted names of the oxoacids (whether known in the free state or not) and some of their peroxo- and thio-derivatives (**5.22** and **5.23**).

For many of these acids systematic names are preferable, and especially for the less common ones, for example:

H_2MnO_4	tetraoxomanganic(VI) acid, to distinguish it from H_3MnO_4, tetraoxomanganic(V) acid.
$HReO_4$	tetraoxorhenic(VII) acid, to distinguish it from H_3ReO_5, pentaoxorhenic(VII) acid.
H_2ReO_4	tetraoxorhenic(VI) acid, to distinguish it from $HReO_3$, trioxorhenic(V) acid; H_3ReO_4, tetraoxorhenic(V) acid; and $H_4Re_2O_7$, heptaoxodirhenic(V) acid.

Names for oxoacids

H_3BO_3	orthoboric acid or boric acid	H_4SiO_4	orthosilicic acid
		$(H_2SiO_3)_n$	metasilicic acid
$(HBO_2)_n$	metaboric acid	HNO_3	nitric acid
H_2CO_3	carbonic acid	HNO_4	peroxonitric acid
HOCN	cyanic acid	HNO_2	nitrous acid
HNCO	isocyanic acid	HOONO	peroxonitrous acid
HONC	fulminic acid	H_2NO_2	nitroxylic acid

$H_2N_2O_2$	hyponitrous acid	H_2SO_3	sulfurous acid
H_3PO_4	orthophosphoric or phosphoric acid	$H_2S_2O_5$	disulfurous acid
$H_4P_2O_7$	diphosphoric or pyrophosphoric acid	$H_2S_2O_2$	thiosulfurous acid
$(HPO_3)_n$	metaphosphoric acid	$H_2S_2O_4$	dithionous acid
H_3PO_5	peroxomono-phosphoric acid	H_2SO_2	sulfoxylic acid
$H_4P_2O_8$	peroxodiphosphoric acid	$H_2S_xO_6$ ($x = 3,4...$)	polythionic acids
$(HO)_2OP–PO(OH)_2$	hypophosphoric acid or diphosphoric(IV) acid	H_2SeO_4	selenic acid
$(HO)_2P–O–PO(OH)_2$	diphosphoric(III, V) acid	H_2SeO_3	selenious acid
H_2PHO_3	phosphonic acid	H_6TeO_6	orthotelluric acid
$H_2P_2H_2O_5$	diphosphonic acid	H_2CrO_4	chromic acid
HPH_2O_2	phosphinic acid	$H_2Cr_2O_7$	dichromic acid
H_3AsO_4	arsenic acid	$HClO_4$	perchloric acid
H_3AsO_3	arsenious acid	$HClO_3$	chloric acid
$HSb(OH)_6$	hexahydroxo-antimonic acid	$HClO_2$	chlorous acid
H_2SO_4	sulfuric acid	$HClO$	hypochlorous acid
$H_2S_2O_7$	disulfuric acid	$HBrO_4$	perbromic acid
H_2SO_5	peroxomonosulfuric acid	$HBrO_3$	bromic acid
$H_2S_2O_8$	peroxodisulfuric acid	$HBrO_2$	bromous acid
$H_2S_2O_3$	thiosulfuric acid	$HBrO$	hypobromous acid
$H_2S_2O_6$	dithionic acid	H_5IO_6	orthoperiodic acid
		HIO_4	periodic acid
		HIO_3	iodic acid
		HIO	hypoiodous acid
		$HMnO_4$	permanganic acid
		H_2MnO_4	manganic acid
		$HTcO_4$	pertechnetic acid
		H_2TcO_4	technetic acid
		$HReO_4$	perrhenic acid
		H_2ReO_4	rhenic acid

Trivial names should not be given to such acids as HNO, $H_2N_2O_3$, $H_2N_2O_4$, *etc.*, of which salts have been described. The salts are to be designated rationally as oxonitrates(I), trioxodinitrates(II), tetraoxodinitrates (III), *etc.*

The names gallic(III) acid, germanic acid, stannic acid, antimonic acid, bismuthic acid, vanadic acid, niobic acid, tantalic acid, telluric acid, molybdic acid, wolframic acid, and uranic acid may be used for substances with indefinite "water content" and degree of polymerization. The inclusion of the STOCK number in the case of gallic(III) acid serves to distinguish it from the organic acid.

For the names of corresponding anions see **3.22.**

5.22—Peroxoacids

The prefix peroxo-, when used in conjunction with the trivial names of acids, indicates substitution of –O– by –O–O–.

Examples:

1. HNO_4	peroxonitric acid
2. H_3PO_5	peroxomonophosphoric acid
3. $H_4P_2O_8$	peroxodiphosphoric acid
4. H_2SO_5	peroxomonosulfuric acid
5. $H_2S_2O_8$	peroxodisulfuric acid

5.23—Thioacids

Acids derived from oxoacids by replacement of oxygen by sulfur are called *thioacids*.

Examples:

1. $H_2S_2O_2$	thiosulfurous acid
2. $H_2S_2O_3$	thiosulfuric acid
3. HSCN	thiocyanic acid

When more than one oxygen atom can be replaced by sulfur the number of sulfur atoms should generally be indicated:

4. H_3PO_3S	monothiophosphoric acid
5. $H_3PO_2S_2$	dithiophosphoric acid
6. H_2CS_3	trithiocarbonic acid
7. H_3AsS_3	trithioarsenious acid
8. H_3AsS_4	tetrathioarsenic acid

The affixes seleno and telluro may be used in a similar manner.

5.24—Chloroacids, etc.

Acids containing ligands other than oxygen and sulfur are generally designated according to the rules in Section 7.

Examples:

1. $HAuCl_4$	hydrogen tetrachloroaurate(III) or tetrachloroauric(III) acid
2. H_2PtCl_4	hydrogen tetrachloroplatinate(II) or tetrachloroplatinic(II) acid
3. H_2PtCl_6	hydrogen hexachloroplatinate(IV) or hexachloroplatinic(IV) acid
4. $H_4Fe(CN)_6$	hydrogen hexacyanoferrate(II) or hexacyanoferric(II) acid
5. $H[PFHO_2]$	hydrogen fluorohydridodioxophosphate or fluorohydridodioxophosphoric acid
6. HPF_6	hydrogen hexafluorophosphate or hexafluorophosphoric acid
7. H_2SiF_6	hydrogen hexafluorosilicate or hexafluorosilicic acid
8. H_2SnCl_6	hydrogen hexachlorostannate(IV) or hexachlorostannic(IV) acid
9. HBF_4	hydrogen tetrafluoroborate or tetrafluoroboric acid
10. $H[BF_2(OH)_2]$	hydrogen difluorodihydroxoborate or difluorodihydroxoboric acid
11. $H[B(C_6H_5)_4]$	hydrogen tetraphenylborate or tetraphenylboric acid

It is preferable to use names of the type hydrogen tetrachloroaurate(III), *etc.* rather than the 'acid' names.

For some of the more important acids of this type abbreviated names may be used, *e.g.*, fluorosilicic acid.

5.3. FUNCTIONAL DERIVATIVES OF ACIDS

Functional derivatives of acids are compounds formed from acids by substitution of OH and sometimes also O by other groups. In this field functional nomenclature is still used but is not recommended.

5.31—Acid Halogenides

The names of acid halogenides are formed from the name of the corresponding acid radical if this has a special name, *e.g.*, nitrosyl chloride, phosphoryl chloride.

In other cases these compounds are named as halogenide oxides according to rule **6.41**, *e.g.*, $MoCl_2O_2$, molybdenum dichloride dioxide.

5.32—Acid Anhydrides

Anhydrides of inorganic acids should generally be given names as oxides, *e.g.*, N_2O_5 dinitrogen pentaoxide, *not* nitric anhydride or nitric acid anhydride.

5.33—Esters

Esters of inorganic acids are given names in the same way as the salts, *e.g.*, dimethyl sulfate, diethyl hydrogenphosphate, trimethyl phosphite.

If, however, it is desired to specify the constitution of the compound, a name according to the nomenclature for coordination compounds should be used.

Examples:

1. $(CH_3)_4\ Fe(CN)_6$	tetramethyl hexacyanoferrate(II)
or	or
2. $[Fe(CN)_2(CNCH_3)_4]$	dicyanotetrakis(methyl isocyanide)iron(II)

5.34—Amides

The names for amides may be derived from the names of acids by replacing acid by amide, or from the names of the acid radicals.

Examples:

1. $SO_2(NH_2)_2$	sulfuric diamide or sulfonyl diamide
2. $PO(NH_2)_3$	phosphoric triamide or phosphoryl triamide

If not all hydroxyl groups of the acid have been replaced by NH_2 groups, names ending in -amidic acid may be used: this is an alternative to naming the compounds as complexes.

Examples:

3. NH_2SO_3H	amidosulfuric acid or sulfamidic acid
4. $NH_2PO(OH)_2$	amidophosphoric acid or phosphoramidic acid
5. $(NH_2)_2PO(OH)$	diamidophosphoric acid or phosphorodiamidic acid

Abbreviated names (sulfamide, phosphamide, sulfamic acid) are often used but are not recommended.

5.35—Nitriles

The suffix -nitrile has been used in the names of a few inorganic compounds, *e.g.*, $(PCl_2N)_3$ trimeric phosphonitrilechloride, but these should be named systematically.

Examples:

1. $(PCl_2N)_3$	trimeric phosphorus dichloride nitride and not trimeric phosphonitrilechloride (*cf.* **2.22**)
2. $K\ [Os^{VIII}(N)O_3]$	potassium nitridotrioxoosmate(VIII) or potassium nitridotrioxoosmate(1–) and not potassium nitriloosmate (*cf.* **7.31**)

There is no reason for retention of the name nitrile (and nitrilo, *cf.* **3.33**) in purely inorganic names.

6. SALTS AND SALT-LIKE COMPOUNDS

Among salts particularly there persist many old names which are bad and misleading, and the Commission wishes to emphasize that any which do not conform to these Rules should be discarded.

6.1. SIMPLE SALTS

Simple salts fall under the broad definition of binary compounds given in Section **2**, and their names are formed from those of the constituent ions (given in Section **3**) in the manner set out in Section **2.**

6.2. SALTS CONTAINING ACID HYDROGEN
("Acid" salts*)

Names are formed by adding the word 'hydrogen', with numerical prefix where necessary, to denote the replaceable hydrogen in the salt. Hydrogen shall be followed without space by the name of the anion. Exceptionally, inorganic anions may contain hydrogen which is not replaceable. It is still designated by hydrogen, if it is considered to have the oxidation number +I, but the salts cannot of course be called acid salts.

Examples:

1. $NaHCO_3$ sodium hydrogencarbonate
2. LiH_2PO_4 lithium dihydrogenphosphate
3. KHS potassium hydrogensulfide
4. $NaHPHO_3$ sodium hydrogenphosphonate

6.3. DOUBLE, TRIPLE, ETC., SALTS

6.31—In formulae all the cations shall precede the anions; in names the following rules shall be applied. In those languages where cation names are placed after anion names the adjectives double, triple, *etc.* (their equivalents in the language concerned) may be added immediately after the anion name. The number so implied concerns the number of *kinds* of cation present and *not* the total number of such ions.

6.32—Cations

6.321—The cations other than hydrogen (*cf.* **6.2** and **6.323**) are to be cited in alphabetical order which may be different in formulae and names.

6.322—Hydration of cations. Owing to the prevalence of hydrated cations, many of which are in reality complexes, it is unnecessary to disturb the cation order to allow for this. If it is necessary to draw attention specifically to the presence of a particular hydrated cation this is treated as a complex ion, *e.g.*, hexaaquazinc, and takes its place in the alphabetical sequence.

6.323—Acidic hydrogen. Hydrogen is cited last among the cations.

* For "basic" salts, see **6.4**.

Examples:

1. $KMgF_3$	magnesium potassium fluoride
2. $NaTl(NO_3)_2$	sodium thallium(I) nitrate or sodium thallium dinitrate
3. $KNaCO_3$	potassium sodium carbonate
4. $MgNH_4PO_4 \cdot 6H_2O$	ammonium magnesium phosphate hexahydrate
5. $Na(UO_2)_3Zn(C_2H_3O_2)_9 \cdot 6H_2O$	sodium triuranyl(VI) zinc nonaacetate hexahydrate
6. $Na(UO_2)_3\,[Zn(H_2O)_6]\,(C_2H_3O_2)_9$	hexaaquazinc sodium triuranyl(VI) nonaacetate
7. $NaNH_4HPO_4 \cdot 4H_2O$	ammonium sodium hydrogenphosphate tetrahydrate
8. $AlK(SO_4)_2 \cdot 12H_2O$	aluminium potassium sulfate 12-water or aluminium potassium bissulfate 12-water

6.33—Anions

Anions are to be cited in alphabetical order which may be different in formulae and names.

6.34—The stoicheiometric method is used for indicating the proportions of constituents if necessary.

Examples:

1. $NaCl{\cdot}NaF{\cdot}2Na_2SO_4$ or $Na_6ClF(SO_4)_2$	hexasodium chloride fluoride bis(sulfate)
2. $Ca_5F(PO_4)_3$	pentacalcium fluoride tris(phosphate)

The multiplicative numerical prefixes bis, tris, *etc.*, are used in connection with the above anions, because di, tri, *etc.*, have been pre-empted to designate condensed anions.

6.4. OXIDE AND HYDROXIDE SALTS
("Basic" salts, formerly oxy- and hydroxy-salts)

6.41—For the purposes of nomenclature, these should be regarded as double salts containing O^{2-} and HO^- anions, and Section **6.3** may be applied in its entirety.

6.42—Use of Prefixes Oxy- and Hydroxy-

In some languages the citation in full of all the separate anion names presents no trouble and is strongly recommended (*e.g.*, copper chloride oxide), to the exclusion of the oxy-form wherever possible. In some other languages, however, such names as "chlorure et oxyde double de cuivre" are so far removed from current practice that the present system of using oxy- and hydroxy-, *e.g.*, oxychlorure de cuivre, may be retained.

Examples:

1. $MgCl(OH)$	magnesium chloride hydroxide
2. $BiClO$	bismuth chloride oxide
3. $LaFO$	lanthanum fluoride oxide
4. $VO(SO_4)$	vanadium(IV) oxide sulfate
5. $CuCl_2{\cdot}3Cu(OH)_2$ or $Cu_2Cl(OH)_3$	dicopper chloride trihydroxide
6. $ZrCl_2O{\cdot}8H_2O$	zirconium dichloride oxide octahydrate

6.5. DOUBLE OXIDES AND HYDROXIDES

The terms "mixed oxides" and "mixed hydroxides" are not recommended. Such substances should preferably be named double, triple, *etc.*, oxides or hydroxides as the case may be.

Many double oxides and hydroxides belong to several distinct groups, each having its own characteristic structure-type which is sometimes named after some well-known mineral of the same group (*e.g.*, perovskite, ilmenite, spinel, *etc.*). Thus, $NaNbO_3$, $CaTiO_3$, $CaCrO_3$, $CuSnO_3$, $YAlO_3$, $LaAlO_3$, and $LaGaO_3$ all have the same structure as perovskite, $CaTiO_3$.* Names such as calcium titanate may convey false implications and it is preferable to name such compounds as double oxides and double hydroxides unless there is clear and generally accepted evidence of cations and oxo- or hydroxo-anions in the structure. This does not mean that names such as titanates or aluminates should always be abandoned, because such substances may exist in solution and in the solid state (*cf.* **3.223**).

6.51—In the double oxides and hydroxides the metals are cited in alphabetical order.

Examples:

1. $Al_2Ca_4O_7 \cdot nH_2O$	dialuminium tetracalcium heptaoxide hydrate
or	
$AlCa_2(OH)_7 \cdot nH_2O$	aluminium dicalcium heptahydroxide hydrate
but	
$Ca_3[Al(OH)_6]_2$	tricalcium bis(hexahydroxoaluminate)
2. $AlLiMn^{IV}{}_2O_4(OH)_4$	aluminium lithium dimanganese(IV) tetrahydroxide tetraoxide

6.52—When required the structure type may be added in parentheses and in italics after the name. When the type-name is also the mineral name of the substance itself the italics should not be used (*cf.* **9.12**). When the structure type is added, the formula and name should be in accordance with the structure.

Examples:

1. $MgTiO_3$	magnesium titanium trioxide (*ilmenite* type)
2. $FeTiO_3$	iron(II) titanium trioxide (ilmenite)
3. $NaNbO_3$	sodium niobium trioxide (*perovskite* type)

* Deviations from alphabetical order in formulae are allowed when compounds with analogous structures are compared (*cf.* **2.17**).

7. COORDINATION COMPOUNDS

7.1. DEFINITIONS

In the oldest sense, the term *coordination entity* generally refers to molecules or ions in which there is an atom (A) to which are attached other atoms (B) or groups (C) to a number in excess of that corresponding to the classical or stoicheiometric valency of the atom (A). However, the system of nomenclature originally evolved for these compounds within this narrow definition has proved useful for a much wider class of compounds, and for the purposes of nomenclature the restriction "in excess of . . . stoicheiometric valency" is to be omitted. Any compound formed by addition of one or several ions and/or molecules to one or more ions and/or molecules may be named according to the same system as strict coordination compounds.

The effect of this definition is to bring many simple and well-known compounds under the same nomenclature rules as those accepted for coordination compounds. Thus, the diversity of names is reduced and also many controversial issues avoided. It is not intended to imply the existence of any constitutional analogy between different compounds merely because they are named under a common system of nomenclature. The system may be extended to many addition compounds.

In the rules which follow certain terms are used in the senses here indicated: the atom referred to above as (A) is known as the *nuclear* or *central* atom, and all other atoms which are directly attached to (A) are known as *coordinating* or *ligating* atoms. Each central atom (A) has a characteristic *coordination number* or *ligancy* which is the number of atoms directly attached to it. Atoms (B) and groups (C) are called *ligands.* A group containing more than one potential coordinating atom is termed a *multidentate* ligand, the number of potential coordinating atoms being indicated by the terms *unidentate, bidentate, etc.* A *chelate* ligand is a ligand attached to one central atom through two or more coordinating atoms, whereas a *bridging group* is attached to more than one centre of coordination. The whole assembly of one or more central atoms with their attached ligands is referred to as a *coordination entity,* which may be a cation, an anion or an uncharged molecule. A *polynuclear* entity is one which contains more than one nuclear atom, their number being designated by the terms *mononuclear, dinuclear, etc.*

Note: Many are puzzled by the use of two different sets of numerical prefixes: uni-, bi-, ter-, and multi-dentate and mono-, di-, tri-, and poly-nuclear. Consistency dictates the use of the Latin prefixes with words of Latin origin: uni, bi, tri (ter), quadri, quinque, sexi, septi, octa; but the Greek prefixes with words of Greek origin: mono, di, tri, tetra, penta, hexa, hepta, octa. In practice, this distinction is not always maintained.

7.2. FORMULAE AND NAMES FOR COORDINATION COMPOUNDS IN GENERAL

7.21—Central Atoms

In *formulae* the usual practice is to place the symbol for the central atom(s) *first* (except in formulae which are primarily structural), with the ionic and neutral ligands following, and the formula for the whole complex entity enclosed in square brackets [ ].*† Order within each class should be in the alphabetical order of the symbols for the ligating atom.

In *names* the central atom(s) should be placed after the ligands. Two kinds of multiplying prefixes are used within the complete name of the coordinating entity (see Preamble): the simple di, tri, *etc.*, derived from the cardinal Greek numerals with simple expressions, and the multiplicative bis, tris, tetrakis, *etc.*, derived from the adverbial forms of Greek numerals with complex expressions or to avoid ambiguity. Enclosing marks are usually used with multiplying numerical prefixes (**2.251**) just as in organic nomenclature. (see *I.U.P.A.C. Nomenclature of Organic Chemistry*, Butterworths, London, 1971, pp. 80–81.) The nesting order for enclosing marks in names is {[( )]}.

7.22—Indication of Oxidation Number and Proportion of Constituents

The names of coordination entities always have been intended to indicate the charge of the central atom (ion) from which the entity is derived. Since the charge on the coordination entity is the algebraic sum of the charges of the constituents, the necessary information may be supplied by giving either the STOCK number (formal charge on the central ion, *i.e.* oxidation number) (see Preamble) or the EWENS-BASSETT number (charge on the entire ion) (**2.252**). [In using EWENS–BASSETT numbers, zero is omitted.] Alternatively the proportion of constituents may be given by means of stoicheiometric prefixes (**2.251**).

Examples:

1. $K_3[Fe(CN)_6]$ potassium hexacyanoferrate(III)
potassium hexacyanoferrate(3–)
tripotassium hexacyanoferrate
2. $K_4[Fe(CN)_6]$ potassium hexacyanoferrate(II)
potassium hexacyanoferrate(4–)
tetrapotassium hexacyanoferrate

7.23—Structural Prefixes

Structural information may be given in formulae and names by prefixes such as *cis*, *trans*, *fac*, *mer*, *etc.* (see **2.19, 7.5, 7.61, 7.62, 7.72,** and Table III).

7.24—Terminations

Anions are given the terminations -ide, -ite, or -ate (*cf.* **2.23**, **2.24**, and **3.223**). Cations and neutral molecules are not given any distinguishing termination. For further details concerning the names of ligands see **7.3**.

* Accordingly in such a case as the ethylene derivative of $PtCl_2$ where the true formula is twice the empirical formula, the complex should be written $[\{PtCl_2(C_2H_4)\}_2]$, not $[PtCl_2(C_2H_4)]_2$ or $[Pt_2Cl_4(C_2H_4)_2]$.

† Enclosing marks are nested within the square brackets as follows:
[( )], [{( )}], [{[( )]}], [{{[( )]}}], *etc.* (*cf.* Preamble).

7.25—Order of Citation of Ligands in Coordination Entities

The ligands are listed in alphabetical order regardless of the number of each. The name of a ligand is treated as a unit. Thus, "diammine" is listed under "a" and "dimethylamine" under "d".

7.3. NAMES FOR LIGANDS

7.31—Anionic Ligands*

7.311—The names for anionic ligands, whether inorganic or organic, end in -o (see, however, **7.313**). In general, if the anion name ends in -ide, -ite, or -ate, the final -e is replaced by -o, giving -ido, -ito and -ato respectively (see also **7.314**). Enclosing marks are required for inorganic anionic ligands containing numerical prefixes, as (triphosphato), and for thio, seleno and telluro analogues of oxo anions containing more than one atom, as (thiosulfato).

Examples of organic anionic ligands which are named in this fashion:

CH_3COO^-	acetato
CH_3OSOO^-	methyl sulfito
$(CH_3)_2N^-$	dimethylamido
CH_3CONH^-	acetamido

7.312—The anions listed below do not follow exactly **7.311**; modified forms have become established, in some cases along with the regular forms:

	ion	*ligand*
F^-	fluoride	fluoro
Cl^-	chloride	chloro
Br^-	bromide	bromo
I^-	iodide	iodo
O^{2-}	oxide	oxo
H^-	hydride	hydrido or hydro†
OH^-	hydroxide	hydroxo
O_2^{2-}	peroxide	peroxo‡
HO_2^-	hydrogenperoxide	hydrogenperoxo
S^{2-}	sulfide	thio
(but: S_2^{2-}	disulfide	disulfido)
HS^-	hydrogen sulfide	mercapto
CN^-	cyanide	cyano
CH_3O^-	methoxide or methanolate	methoxo‡ or methanolato
CH_3S^-	methanethiolate	methylthio or methanethiolato

Examples:

The letter in each of the ligand names which is used to determine the alphabetical listing is given in bold face type in the following examples to illustrate the alphabetical arrangement.§ For many compounds, the oxidation number of the central atom and/or the charge on the ion are so well known that there is no need to use either a STOCK

* As a general term for anionic ligands, "aniono" may be used, *e.g.*, tetraamminedianionocobalt(III) ion. The term "acido" is discouraged.

† Both hydrido and hydro are used for coordinated hydrogen but the latter term usually is restricted to boron compounds.

‡ In conformity with the practice of organic nomenclature, the forms peroxy and methoxy are also used but are not recommended.

§ The use of bold face type is not a nomenclature rule.

number or a EWENS–BASSETT number. However, it is not wrong to use such numbers and they are included here.

1. Na $[B(NO_3)_4]$
   sodium tetranitratoborate(1–)
   sodium tetranitratoborate(III)
2. K_2 $[OsCl_5N]$
   potassium pentachloronitridoosmate(2–)
   potassium pentachloronitridoosmate(VI)
3. $[Co(NH_2)_2(NH_3)_4]$ OC_2H_5
   diamidotetraamminecobalt(1+) ethoxide
   diamidotetraamminecobalt(III) ethoxide
4. $[CoN_3(NH_3)_5]$ SO_4
   pentaammineazidocobalt(2+) sulfate
   pentaammineazidocobalt(III) sulfate
5. Na_3 $[Ag(S_2O_3)_2]$
   sodium bis(thiosulfato)argentate(3–)
   sodium bis(thiosulfato)argentate(I)
6. $[Ru(HSO_3)_2(NH_3)_4]$
   tetraamminebis(hydrogensulfito)ruthenium
   tetraamminebis(hydrogensulfito)ruthenium(II)
7. NH_4 $[Cr(NCS)_4(NH_3)_2]$
   ammonium diamminetetrakis(isothiocyanato)chromate(1–)
   ammonium diamminetetrakis(isothiocyanato)chromate(III)
8. K $[AgF_4]$
   potassium tetrafluoroargentate(1–)
   potassium tetrafluoroargentate(III)
9. Ba $[BrF_4]_2$
   barium tetrafluorobromate(1–)
   barium tetrafluorobromate(III)
10. Cs $[ICl_4]$
    caesium tetrachloroiodate(1–)
    caesium tetrachloroiodate(III)
11. K $[Au(OH)_4]$
    potassium tetrahydroxoaurate(1–)
    potassium tetrahydroxoaurate(III)
12. K $[CrF_4O]$
    potassium tetrafluorooxochromate(1–)
    potassium tetrafluorooxochromate(V)
13. K_2 $[Cr(CN)_2O_2(O_2)NH_3]$
    potassium amminedicyanodioxoperoxochromate(2–)
    potassium amminedicyanodioxoperoxochromate(VI)
14. $[AsS_4]^{3-}$
    tetrathioarsenate(3–) ion
    tetrathioarsenate(V) ion
15. K_2 $[Fe_2(NO)_4S_2]$*
    potassium tetranitrosyldithiodiferrate(2–)
16. K $[AuS(S_2)]$
    potassium (disulfido)thioaurate(1–)
    potassium (disulfido)thioaurate(III)

There is no elision of vowels or use of a diaeresis in tetraammine and similar names.

7.313—Although the common hydrocarbon radicals generally behave as anions when they are attached to metals and in fact are sometimes encountered as anions, their presence in coordination entities is indicated by the customary radical names even though they are considered as anions in computing the oxidation number.

* The name with a STOCK number is omitted because there is no general agreement on what the STOCK number should be.

Examples:

1. $K[B(C_6H_5)_4]$
   potassium tetraphenylborate(1−)
   potassium tetraphenylborate(III)
2. $K[SbCl_5(C_6H_5)]$
   potassium pentachloro(phenyl)antimonate(1−)*
   potassium pentachloro(phenyl)antimonate(V)
3. $K_2[Cu(C_2H)_3]$
   potassium triethynylcuprate(2−)
   potassium triethynylcuprate(I)
4. $K_4[Ni(C_2C_6H_5)_4]$
   potassium tetrakis(phenylethynyl)niccolate(4−)
   potassium tetrakis(phenylethynyl)niccolate(0)
5. $[Fe(C_2C_6H_5)_2(CO)_4]$
   tetracarbonylbis(phenylethynyl)iron
   tetracarbonylbis(phenylethynyl)iron(II)

7.314—Ligands derived from organic compounds by the loss of protons (other than those named by **7.311, 7.312**, and **7.313**) are given the ending -ato. Enclosing marks are used to set off all such ligand names regardless of whether they are substituted or unsubstituted, *e.g.*, (benzoato), (*p*-chlorophenolato), [2-(chloromethyl)-1-naphtholato]. (If the ligand coordinates with no loss of a proton, the organic compound is used without alteration—**7.32**). Where a neutral organic compound forms ligands with different charges by the loss of different numbers of protons, the charge shall be designated in parentheses after the name of the ligand; *e.g.*, $^{-}OOCCH(O^{-})CH(OH)COO^{-}$ is tartrato(3–) and $^{-}OOCCH(OH)CH(OH)COO^{-}$ is tartrato(2–). The use of EWENS–BASSETT numbers as part of the ligand name requires the use of square brackets around the ligand names and the multiplicative prefixes bis, tris, *etc.*, rather than di, tri, *etc.* Inorganic chemists have held to trivial and older forms of names for many of the common ligands: cupferron, dithizone, 8-hydroxyquinoline or oxine, acetylacetone, dipyridyl, tripyridyl, *etc.* instead of *N*-nitroso-*N*-phenylhydroxylamine, 1,5-diphenylthiocarbazone, 8-quinolinol, 2,4-pentanedione, 2,2′-bipyridine (bipyridyl), 2,2′ : 6′,2″-terpyridine, *etc.* In the interest of a uniform nomenclature among organic and coordination compounds, the names used here for organic ligands are in accord with the Definitive Rules for the Nomenclature of Organic Chemistry (IUPAC Commission on the Nomenclature of Organic Chemistry).

Examples:

1. $[Ni(C_4H_7N_2O_2)_2]$ — bis(2,3-butanedione dioximato)nickel
   bis(2,3-butanedione dioximato)nickel(II)
2. $[Cu(C_5H_7O_2)_2]$ — bis(2,4-pentanedionato)copper
   bis(2,4-pentanedionato)copper(II)
3. 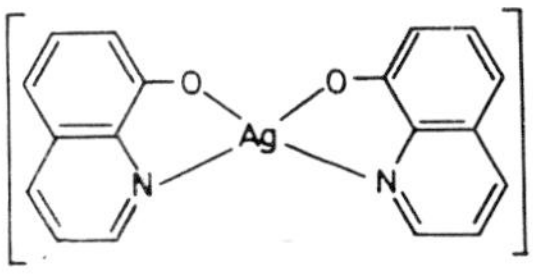 [structure: O, Ag, N, O, N] — bis(8-quinolinolato)silver
   bis(8-quinolinolato)silver(II)

* Normally phenyl would not be placed within enclosing marks. They are used here to avoid confusion with a chlorophenyl radical.

4. 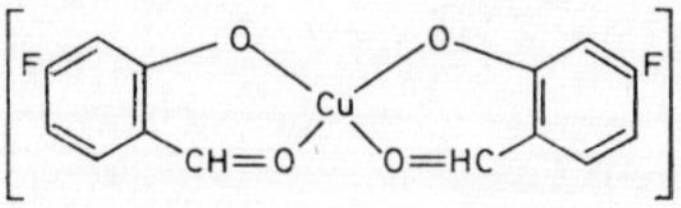 bis(4-fluorosalicylaldehydato)copper
bis(4-fluorosalicylaldehydato)copper(II)

5. 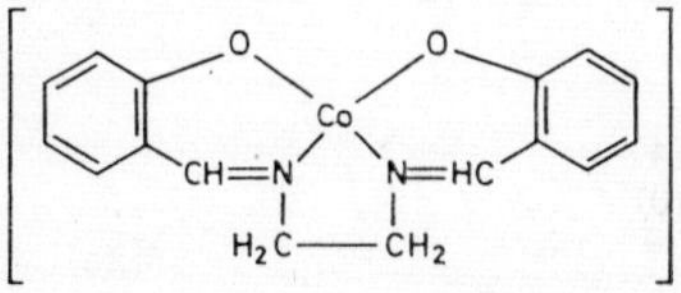 *N,N′*-ethylenebis(salicylideneiminato)cobalt
N,N′-ethylenebis(salicylideneiminato)cobalt(II)

7.32—Neutral and Cationic Ligands

7.321—The name of a coordinated molecule is to be used without change, except for the special cases provided for in **7.322** and **7.323**. All neutral ligands other than those in **7.322** and **7.323** are set off with enclosing marks.

Examples:

1. $[CoCl_2(C_4H_8N_2O_2)_2]$
bis(2,3-butanedione dioxime)dichlorocobalt
bis(2,3-butanedione dioxime)dichlorocobalt(II) (*cf.* example 1 in **7.314**)
2. *cis*-$[PtCl_2(Et_3P)_2]$
cis-dichlorobis(triethylphosphine)platinum
cis-dichlorobis(triethylphosphine)platinum(II)
3. $[CuCl_2(CH_3NH_2)_2]$
dichlorobis(methylamine)copper
dichlorobis(methylamine)copper(II)
4. $[Pt(py)_4]$ $[PtCl_4]$
tetrakis(pyridine)platinum(2+) tetrachloroplatinate(2−)
tetrakis(pyridine)platinum(II) tetrachloroplatinate(II)
5. $[Fe(bpy)_3]$ Cl_2
tris(2,2′-bipyridine)iron(2+) chloride
tris(2,2′-bipyridine)iron(II) chloride
6. $[Co(en)_3]_2$ $(SO_4)_3$
tris(ethylenediamine)cobalt(3+) sulfate
tris(ethylenediamine)cobalt(III) sulfate
7. $[Zn\{NH_2CH_2CH(NH_2)CH_2NH_2\}_2]$ I_2
bis(1,2,3-propanetriamine)zinc(2+) iodide
bis(1,2,3-propanetriamine)zinc(II) iodide
8. K $[PtCl_3(C_2H_4)]$
potassium trichloro(ethylene)platinate(1−)
potassium trichloro(ethylene)platinate(II) or
potassium trichloromonoethyleneplatinate(II)
9. $[Cr(C_6H_5NC)_6]$
hexakis(phenyl isocyanide)chromium
hexakis(phenyl isocyanide)chromium(0)
10. $[Ru(NH_3)_5(N_2)]$ Cl_2
pentaammine(dinitrogen)ruthenium(2+) chloride
pentaammine(dinitrogen)ruthenium(II) chloride
11. $[CoH(N_2)\{(C_6H_5)_3P\}_3]$
(dinitrogen)hydridotris(triphenylphosphine)cobalt
(dinitrogen)hydridotris(triphenylphosphine)cobalt(I)

7.322—Water and ammonia as neutral ligands in coordination complexes are called "aqua" (formerly "aquo") and "ammine," respectively.

Examples:

1. $[Cr(H_2O)_6]\,Cl_3$
   hexaaquachromium(3+) chloride
   hexaaquachromium trichloride
2. $[Al(OH)(H_2O)_5]^{2+}$
   pentaaquahydroxoaluminium(2+) ion
   pentaaquahydroxoaluminium(III) ion
3. $[Co(NH_3)_6]\,Cl(SO_4)$
   hexaamminecobalt(3+) chloride sulfate
   hexaamminecobalt(III) chloride sulfate
4. $[CoCl(NH_3)_5]\,Cl_2$
   pentaamminechlorocobalt(2+) chloride
   pentaamminechlorocobalt(III) chloride
5. $[CoCl_3(NH_3)_2\{(CH_3)_2NH\}]$
   diamminetrichloro(dimethylamine)cobalt
   diamminetrichloro(dimethylamine)cobalt(III)

7.323—The groups NO and CO, when linked directly to a metal atom, are called nitrosyl and carbonyl, respectively. In computing the oxidation number these ligands are treated as neutral.

Examples:

1. $Na_2\,[Fe(CN)_5NO]$
   sodium pentacyanonitrosylferrate(2−)
   sodium pentacyanonitrosylferrate(III)
2. $K_3\,[Fe(CN)_5CO]$
   potassium carbonylpentacyanoferrate(3−)
   potassium carbonylpentacyanoferrate(II)
3. $K\,[Co(CN)(CO)_2(NO)]$
   potassium dicarbonylcyanonitrosylcobaltate(1−)
   potassium dicarbonylcyanonitrosylcobaltate(0)
4. $[CoH(CO)_4]$
   tetracarbonylhydridocobalt
   tetracarbonylhydridocobalt(I)
5. $Na\,[Co(CO)_4]$
   sodium tetracarbonylcobaltate(1−)
   sodium tetracarbonylcobaltate(−I)
6. $[Ni(CO)_2(Ph_3P)_2]$
   dicarbonylbis(triphenylphosphine)nickel
   dicarbonylbis(triphenylphosphine)nickel(0)
7. $[Fe(en)_3]\,[Fe(CO)_4]$
   tris(ethylenediamine)iron(2+) tetracarbonylferrate(2−)
   tris(ethylenediamine)iron(II) tetracarbonylferrate(−II)

7.324—The name of a coordinated cation is used without change where there is no ambiguity.

Examples:

1. $[PtCl_2\{H_2NCH_2CH(NH_2)CH_2NH_3\}]\,Cl$
   dichloro(2,3-diaminopropylammonium)platinum(1+) chloride
   dichloro(2,3-diaminopropylammonium)platinum(II) chloride
2. $[NiCl_3(H_2O)\{N(CH_2CH_2)_3NCH_3\}]$
   aquatrichloro{1-methyl-4-aza-1-azoniabicyclo[2.2.2]octane}nickel
   aquatrichloro{1-methyl-4-aza-1-azoniabicyclo[2.2.2]octane}nickel(II)

7.33—Different Modes of Linkage of Some Ligands

The different points of attachment of a ligand may be denoted by adding the italicized symbol(s) for the atom or atoms through which attachment occurs at the end of the name of the ligand. Thus the dithiooxalato anion conceivably may be attached through S or O, and these are distinguished as dithiooxalato-*S*,*S*′ and dithiooxalato-*O*,*O*′, respectively. For the order of citing symbols for unsymmetrical ligands see **7.513(b)**.

In some cases different names are in use for different modes of attachment, for example, thiocyanato (–SCN) and isothiocyanato (–NCS), nitro (–NO_2), and nitrito (–ONO). In the absence of structural knowledge about the linkage actually present, thiocyanato and nitrito should be used.

Examples:

1. $K_2\left[Ni\left(\begin{array}{l}S{-}CO\\ \quad | \\ S{-}CO\end{array}\right)_2\right]$

   potassium bis(dithiooxalato-*S*,*S*′)niccolate(2−)
   potassium bis(dithiooxalato-*S*,*S*′)niccolate(II)

2. $\left[\begin{array}{l}Cl \diagdown \quad\quad NH_2{-}CH_2 \\ \quad\quad Pt \quad\quad\quad | \\ Cl \diagup \quad\quad S\ {-}\ CH_2 \\ \quad\quad\quad\quad | \\ \quad\quad\quad CH_2{-}CH_2N(CH_3)_2\end{array}\right]$

   dichloro[*N*,*N*-dimethyl-2,2′-thiobis(ethylamine)-*S*,*N*′]platinum
   dichloro[*N*,*N*-dimethyl-2,2′-thiobis(ethylamine)-*S*,*N*′]platinum(II)
3. $K_2\ [Pt(NO_2)_4]$
   potassium tetranitroplatinate(2−)
   potassium tetranitroplatinate(II)
4. $Na_3\ [Co(NO_2)_6]$
   sodium hexanitrocobaltate(3−)
   sodium hexanitrocobaltate(III)
5. $[Co(NO_2)_3(NH_3)_3]$ (facial and meridional isomers possible)
   triamminetrinitrocobalt
   triamminetrinitrocobalt(III)
6. $[Co(ONO)(NH_3)_5]\ SO_4$
   pentaamminenitritocobalt(2+) sulfate
   pentaamminenitritocobalt(III) sulfate
7. $[Co(NCS)(NH_3)_5]\ Cl_2$
   pentaammineisothiocyanatocobalt(2+) chloride
   pentaammineisothiocyanatocobalt(III) chloride

There are some ligands which occasionally coordinate in a manner quite different from the normal mode. At times the mode of coordination can be indicated by the names of the ligands; at others, by designating by italicized symbol(s) the atom(s) of the ligand through which coordination occurs.

Examples:

8. $\begin{array}{l}{-}NH_2CH_2\\ \quad\quad\quad | \\ {-}\bar{O}{-}C{=}O\end{array}$ — glycinato-*O*,*N* (where -*O*,*N* is omitted, it is implied)

9. $-NH_2CH_2COOH$ — glycine-*N*

10. $—\bar{O}OCCH_2NH_2$ glycinato-*O*
11. $—\bar{O}OCCH_2\overset{+}{N}H_3$ glycine-*O*
12. $[Pt(CH_3)_3\{CH(COCH_3)_2\}(bpy)]$
 (1-acetylacetonyl)(2,2′-bipyridine)trimethylplatinum or
 (2,2′-bipyridine)(diacetylmethyl)trimethylplatinum or
 (2,2′-bipyridine)trimethyl(2,4-pentanedionato-C^3)platinum
 (for the meaning of -C^3, see **7.34**).

7.34—Designation of Active Coordination Sites from among Several Possibilities

In some cases, several possible coordination sites may be involved. The different possible locations may be indicated by the particular atoms through which coordination occurs: cysteinato-*S,N*; cysteinato-*O,N*; *etc.* If the same element is involved in the different possibilities, the position in the chain or ring to which the element is attached is indicated by numerical superscripts.

Examples:

1.
```
O=C—O              O=CO⁻               O=C—O
  |    \             |                   |     \
  |     M            |                   |      \
  |    /             |                   |       \
 HCO                HCO                 HCOH      \
  |                  |   \               |         M
  |                  |    M              |        /
  |                  |   /               |       /
 HCOH               HCO                 HCOH    /
  |                  |                   |     /
O=C—O⁻             O=C—O⁻              O=C—O
```
tartrato(3−)-O^1,O^2 tartrato(4−)-O^2,O^3 tartrato(2−)-O^1, O^4

2.
```
CH3COCHCOCH3
       |
```
2,4-pentanedionato-C^3-
or 1-acetylacetonyl-
or diacetylmethyl-

7.35—Use of Abbreviations

In the literature of coordination compounds, abbreviations for ligand names are used extensively, especially in formulae. A list of common abbreviations is given in (9) below. Unfortunately, divergent practices have developed with consequent confusion. The following simple rules should govern the use of abbreviations:

1. It should be assumed that the reader will not be familiar with any but the most common abbreviations. Consequently, each paper should explain the abbreviations used in it.
2. Abbreviations should be short—generally not in excess of four letters.
3. Abbreviations should be such as not to cause confusion with the commonly accepted abbreviations used for organic radicals: Me, methyl; Et, ethyl; Ph, phenyl, *etc.*
4. All abbreviations for ligands, except L the general abbreviation for ligand and those shown with initial capital letters under 6, shall be in lower case letters: en, pn, tren, bpy, *etc.*; M shall be the general abbreviation for metal.

5. Abbreviations should not involve hyphens: *e.g.*, phen not *o*-phen for *o*-phenanthroline (or 1,10-phenanthroline).

6. The neutral compound and the ligand ion derived from it should be clearly differentiated.

Hacac	acetylacetone; acac acetylacetonato
H_2dmg	dimethylglyoxime (2,3-butanedione dioxime)
Hdmg	dimethylglyoximato(1–)
dmg	dimethylglyoximato(2–)
H_4edta	ethylenediaminetetraacetic acid
Hedta or edta	coordinated ions derived from H_4edta

7. In using abbreviations, care should be taken to be sure there is no confusion with symbols. The abbreviations should be separated from symbols or enclosed in parentheses.

   $[Co(en)_3]^{3+}$ or $[Co\ en_3]^{3+}$ not $[Coen_3]^{3+}$

8. Abbreviations for molecules or ions combined with the commonly used symbols for organic groups (such as Eten for *N*-ethylethylenediamine; Meacac for 3-methyl-2,4-pentanedione (methylacetylacetone); Etbg for ethylbiguanide) are to be avoided. Simplified formulae or those using symbols for organic groups are preferable.

   $CH_3COCH(C_2H_5)\ COCH_3$ $HCEtAc_2$

9. Commonly used abbreviations are:

Anionic groups (parent acids given)

Hacac	acetylacetone, 2,4-pentanedione, $CH_3COCH_2COCH_3$
Hbg	biguanide, $H_2NC(NH)NHC(NH)NH_2$
H_2dmg	dimethylglyoxime, 2,3-butanedione dioxime, $CH_3C(=NOH)C(=NOH)CH_3$
H_4edta	ethylenediaminetetraacetic acid, $(HOOCCH_2)_2NCH_2CH_2N(CH_2COOH)_2$
H_2ox	oxalic acid, HOOC—COOH

Neutral groups

bpy	2,2′-bipyridine or 2,2′-bipyridyl, 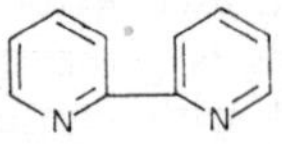
diars	*o*-phenylenebis(dimethylarsine), $(CH_3)_2AsC_6H_4As(CH_3)_2$
dien	diethylenetriamine, $H_2NCH_2CH_2NHCH_2CH_2NH_2$
diphos	ethylenebis(diphenylphosphine), $Ph_2PCH_2CH_2PPh_2$
en	ethylenediamine, $H_2NCH_2CH_2NH_2$
phen	1,10-phenanthroline, 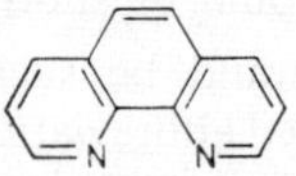
pn	propylenediamine, $H_2NCH(CH_3)CH_2NH_2$
py	pyridine
tren	2,2′, 2″-triaminotriethylamine, $(H_2NCH_2CH_2)_3N$
trien	triethylenetetraamine, $(H_2NCH_2CH_2NHCH_2)_2$
ur	urea, $(H_2N)_2CO$

7.4. COMPLEXES WITH UNSATURATED MOLECULES OR GROUPS

A wide variety of unsaturated hydrocarbon–metal compounds are known. In many of these the metal atom is bonded to two or more contiguous atoms of the ligand rather than to a specific atom. Since the electrons which constitute the π-system in the ligand are involved in the metal–ligand bond, these compounds have been called π-complexes. However, the exact nature of the bonding is often uncertain. Therefore, it seems wise to indicate the atoms bonded to the metal atom in a manner completely independent of theoretical implications. [F. A. Cotton, *J. Amer. Chem. Soc.*, **90**, 6230 (1968)]. Also the attachment of a metal to adjacent atoms in ligands other than unsaturated hydrocarbons occurs: *e.g.*, to C=N–, –N=N–, >N–N<, O=O, $O{-}O^{2-}$.

It is desirable to have a system of nomenclature such that no major changes are required in names when the state of knowledge increases from that of stoicheiometric composition to that of the complete details of the structure. This can be done by application of the following rules and requires only that the ligand can be named and that the ligating atoms can be designated by rules of organic nomenclature.

7.41—Designation of Stoicheiometric Composition Only

The name of the ligand group is given in the usual manner.

Examples:

1. $[PtCl_2(C_2H_4)(NH_3)]$
   amminedichloro(ethylene)platinum
   amminedichloro(ethylene)platinum(II)
2. $K[PtCl_3(C_2H_4)]$
   potassium trichloro(ethylene)platinate(1−)
   potassium trichloro(ethylene)platinate(II)
3. $[Cr(C_6H_6)_2]$
   bis(benzene)chromium
   bis(benzene)chromium(0)
4. $[Ni(C_5H_5)_2]$
   bis(cyclopentadienyl)nickel
   bis(cyclopentadienyl)nickel(II)
5. $[Fe(CO)_3(C_8H_8)]$
   tricarbonyl(cyclooctatetraene)iron
   tricarbonyl(cyclooctatetraene)iron(0)
6. $[Mn(CO)_4\{CH_2{=}C(CH_3)CH_2\}]$
   tetracarbonyl(2-methylallyl)manganese
   tetracarbonyl(2-methylallyl)manganese(I)

7.42—Designation of Structure

7.421—Designation of structure where all the atoms in a chain or ring are bound to the central atom. The name of the ligand group is given as before but with the prefix η. η may be read as eta or hapto (from the Greek *haptein*, ἅπτειν, to fasten).*

Examples:

1. $[PtCl_2(C_2H_4)(NH_3)]$
   amminedichloro(η-ethylene)platinum
   amminedichloro(η-ethylene)platinum(II)

* *cf.* F. A. Cotton, *J. Amer. Chem. Soc.*, **90**, 6230 (1968)

2. $K[PtCl_3(C_2H_4)]$
   potassium trichloro(η-ethylene)platinate(1−)
   potassium trichloro(η-ethylene)platinate(II)
3. $[Cr(C_6H_6)_2]$
   bis(η-benzene)chromium
   bis(η-benzene)chromium(0)
4. $[Ni(C_5H_5)_2]$
   bis(η-cyclopentadienyl)nickel
   bis(η-cyclopentadienyl)nickel(II)
5. $[ReH(C_5H_5)_2]$
   bis(η-cyclopentadienyl)hydridorhenium
   bis(η-cyclopentadienyl)hydridorhenium(III)
6. $[Cr(CO)_3(C_6H_6)]$
   (η-benzene)tricarbonylchromium
   (η-benzene)tricarbonylchromium(0)
7. $[Co(C_5H_5)(C_5H_6)]$
   (η-cyclopentadiene)(η-cyclopentadienyl)cobalt
   (η-cyclopentadiene)(η-cyclopentadienyl)cobalt(I)
8. $[Ni(C_5H_5)(NO)]$
   (η-cyclopentadienyl)nitrosylnickel
9. 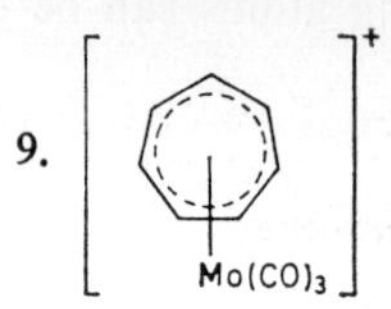

   tricarbonyl(η-cycloheptatrienylium)molybdenum(1+) ion
   tricarbonyl(η-cycloheptatrienylium)molybdenum(0) ion

7.422—Designation of structure where all the multiply-bonded ligand atoms are bound to the central atom. The names are derived as in **7.421**.

Examples:

1. tetracarbonyl(η-1,5-cyclooctadiene)molybdenum
   tetracarbonyl(η-1,5-cyclooctadiene)molybdenum(0)

2. (η-bicyclo[2.2.1]hepta-2,5-diene)tricarbonyliron
   (η-bicyclo[2.2.1]hepta-2,5-diene)tricarbonyliron(0)

7.423—Designation of structure where some, but not all, ligand atoms in a chain or ring or some, but not all, ligand atoms involved in double bonds are bound to the central atom. Locant designators are inserted preceding η. When a number of adjacent atoms in the ligand are in contact with the central atom, the ligand atoms are designated inclusively rather than individually. When it is desired to stress that a ligand is bonded to a single atom, the prefix σ- may be used (*cf.* Example 9).

Examples:

1. 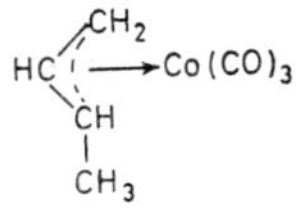 (1–3-η-2-butenyl)tricarbonylcobalt
(1–3-η-2-butenyl)tricarbonylcobalt(I)

2. 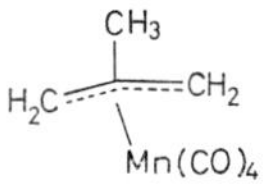 tetracarbonyl[1–3-η-(2-methylallyl)]manganese
tetracarbonyl[1–3-η-(2-methylallyl)]manganese(I)

3. 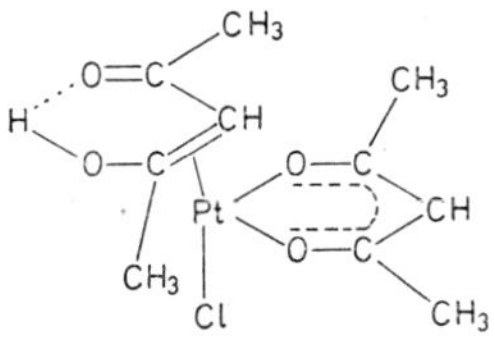 chloro[3-4-η-(4-hydroxy-3-penten-2-one)](2,4-pentanedionato)platinum
chloro[3-4-η-(4-hydroxy-3-penten-2-one)](2,4-pentanedionato)platinum(II)

4. 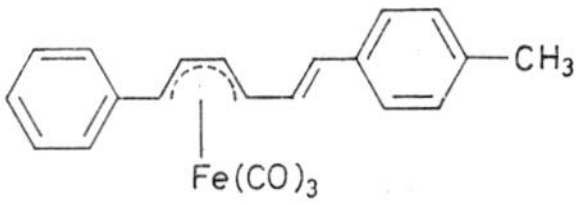 tricarbonyl[1–4-η-(1-phenyl-6-*p*-tolyl-1,3,5-hexatriene)]iron

5. 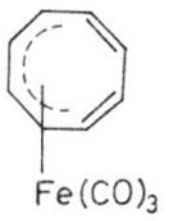 tricarbonyl(1–4-η-cyclooctatetraene)iron

6. 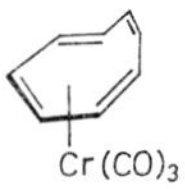 tricarbonyl(1–6-η-cyclooctatetraene)chromium

7. 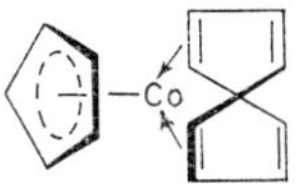 (1-2:5-6-η-cyclooctatetraene)(η-cyclopentadienyl)-cobalt

8. 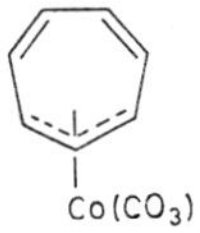 tricarbonyl(1–3-η-cycloheptatrienyl)cobalt

9. 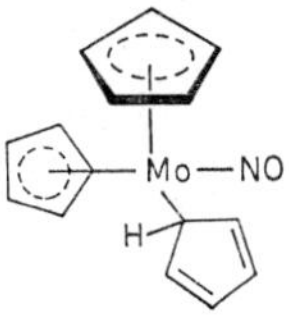 (η-cyclopentadienyl)(1–3-η-cyclopentadienyl)-(σ-cyclopentadienyl)nitrosylmolybdenum

10. dicarbonyl(η-cyclopentadienyl)[α-2-η-(4-methylbenzyl)]molybdenum

11. μ-[1–3a(9a)-η:4–6-η-azulene]-pentacarbonyldiiron-(*Fe-Fe*) (*cf.* **7.71**)

12. *trans*-μ-(1–4-η:5–8-η-cyclooctatetraene)-bis-(tricarbonyliron) (*cf.* **7.51, 7.61**)

7.43—Cyclopentadienyl Complexes: Metallocenes

The general term for η-cyclopentadienyl complexes and their derivatives is metallocenes. The trivial name for bis(η-cyclopentadienyl)iron, $Fe(C_5H_5)_2$, is ferrocene. The term "sandwich compounds" is too general to be used to designate the metallocenes specifically. "Ocene" names (nickelocene, cobaltocene, osmocene, *etc.*) should not be employed for individual compounds (at least, as long as their organic chemistry is not as well developed as that of ferrocene). The introduction of other trivial names, such as cymantrene and cytizel should be avoided.

Examples:

1. $Fe(C_5H_5)_2$
   bis(η-cyclopentadienyl)iron
   bis(η-cyclopentadienyl)iron(II)
   ferrocene
2. $[Fe(C_5H_5)_2]$ $[BF_4]$
   bis(η-cyclopentadienyl)iron(1+) tetrafluoroborate
   bis(η-cyclopentadienyl)iron(III) tetrafluoroborate
   ferrocene(1+) tetrafluoroborate(1−)
   ferrocenium tetrafluoroborate

7.431—Derivatives of Ferrocene. Derivatives of ferrocene are named by the use of the prefixes and suffixes for organic substituents. However, any substituent may be indicated by a prefix (*cf.* **7.432**). Since all of the positions on the cyclopentadiene rings may be considered equivalent, substituents are given low numbering without regard to attachment of the iron. The second cyclopentadiene ring is numbered with primed numbers: 1′, 2′, *etc.* If the compound structure contains two ferrocene groups, doubly and triply primed numbers (1″, 1‴, *etc.*) are used for the third and fourth cyclopentadienyl groups.

Examples:

1. 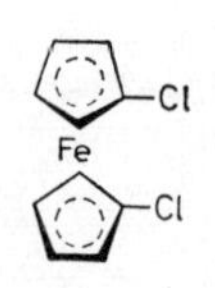

1,1′-dichloroferrocene

2. 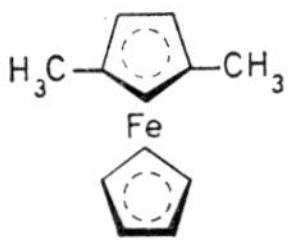 1,3-dimethylferrocene

3. 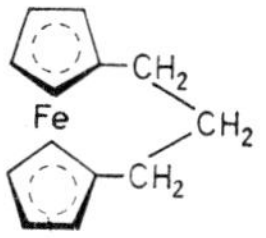 1,1′-trimethyleneferrocene
1,3-(1,1′-ferrocenediyl)propane

4. 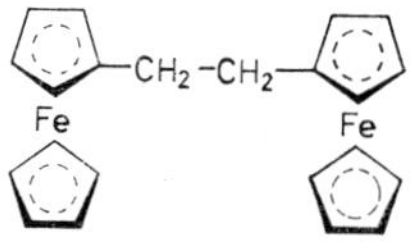 1,1″-ethylenediferrocene

5. 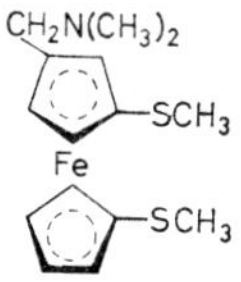 3-[(dimethylamino)methyl]-1,1′-bis(methylthio)ferrocene or *N*,*N*-dimethyl-1′,3-bis(methylthio)-1-ferrocenemethylamine (*cf.* **7.432**)

6. $[Fe(C_2H_5C_5H_4)(C_5H_5)]Cl$ ethylferrocene(1+) chloride or ethylferrocenium chloride

7.432—Ferrocenyl Radicals (in part alternative to **7.431**). Ferrocene derivatives containing a principal group which can be designated by a suffix may alternatively be named as an organic parent compound with a ferrocene radical as substituent. When necessary, radical names such as ferrocenyl, ferrocenediyl, and ferrocenetriyl are used.

Examples:

1. $(C_{10}H_9Fe)$—$COCH_3$	ferrocenyl methyl ketone or acetylferrocene
2. $(C_{10}H_9Fe)$—CHO	ferrocenecarbaldehyde or formylferrocene
3. $(C_{10}H_9Fe)$—CH_2OH	ferrocenylmethanol or (hydroxymethyl)ferrocene
4. $(C_{10}H_9Fe)$—$COOH$	ferrocenecarboxylic acid or carboxyferrocene
5. $(C_{10}H_9Fe)$—CH_2CHNH_2COOH	α-aminoferrocenepropionic acid or 3-ferrocenylalanine
6. $(C_{10}H_9Fe)$—$As(C_6H_5)_2$	ferrocenyldiphenylarsine or (diphenylarsino)ferrocene
7. $(C_{10}H_9Fe)_2NC_2H_5$	*N*-ethyl-1,1″-diferrocenylamine or 1,1″-(ethylimino)diferrocene
8. $(C_{10}H_9Fe)$—$\overset{+}{N}(CH_3)_3$	ferrocenyltrimethylammonium or (trimethylammonio)ferrocene

9. 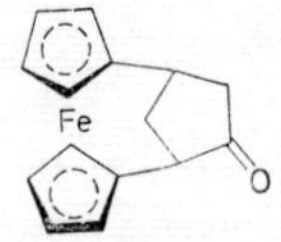 2,4-(1,1′-ferrocenediyl)cyclopentanone

The above treatment is not extended to ferrocene derivatives containing other rings fused to the cyclopentadiene ring.

10. 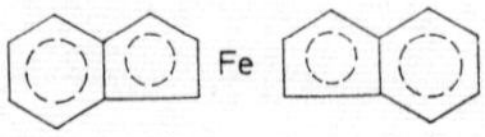 bis[1–3a(7a)-η-indenyl]iron (not benzoferrocene or dibenzoferrocene)

11. 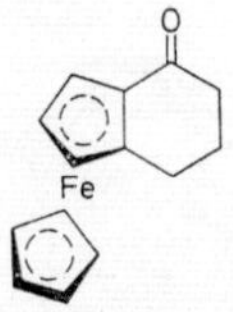 (η-cyclopentadienyl)[1–3a(7a)-η-4,5,6,7-tetrahydro-4-oxoindenyl]iron
(η-cyclopentadienyl)[1–3a(7a)-η-4,5,6,7-tetrahydro-4-oxoindenyl]iron(II)

7.433—Absolute Configurations. The absolute configuration of enantiomers is specified by the sequence-rule method.*

Examples:

1. 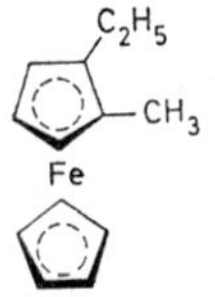 (1*S*)-1-ethyl-2-methylferrocene

2. 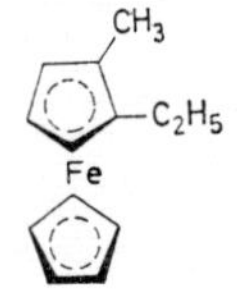 (1*R*)-1-ethyl-2-methylferrocene

7.434—If in the crystalline state a ferrocene derivative has a preferred conformation, the conformations are described in the manner of **E–6.6** of Nomenclature of Organic Chemistry, IUPAC 1968 Tentative Rules, Section E, *IUPAC Information Bulletin* No. 35, June 1969, p. 65.

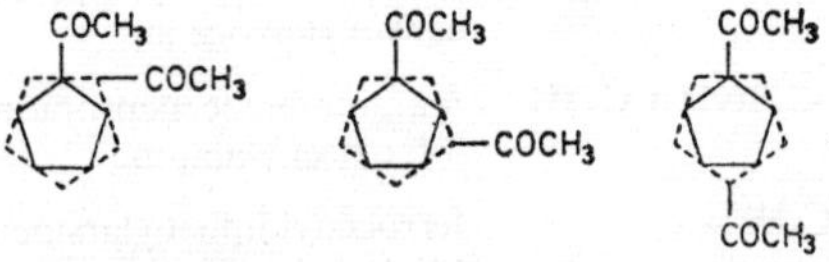

synclinal(sc) anticlinal(ac) antiperiplanar(ap)

* Nomenclature of Organic Chemistry, IUPAC 1968 Tentative Rules, Section E, *IUPAC Information Bull.* No. 35, June 1969, pp. 71–79; *cf.* 'Convention for π-complexes' by R. S. Cahn, Sir Christopher Ingold, and V. Prelog, *Angew. Chem. Internat. Edn.*, **5**, 394 (1966).

7.5. DESIGNATION OF ISOMERS

Among coordination compounds, isomerism may arise in a number of ways:

(a) Different atoms of the ligand through which coordination to a central atom occurs—**7.33; 7.34.**

(b) Coordination of isomeric ligands—indicated by the name of the ligand:

$H_2NCH(CH_3)CH_2NH_2$	1,2-propanediamine
$CH_3NHCH_2CH_2NH_2$	*N*-methylethylenediamine

(c) Interchange of ions between coordination sphere and ionic sphere—indicated by the names:

$[CoSO_4(NH_3)_5]\,Br$
pentaamminesulfatocobalt(III) bromide
$[CoBr(NH_3)_5]\,SO_4$
pentaamminebromocobalt(III) sulfate

(d) Geometrical arrangement of two or more kinds of ligands in the coordination sphere:

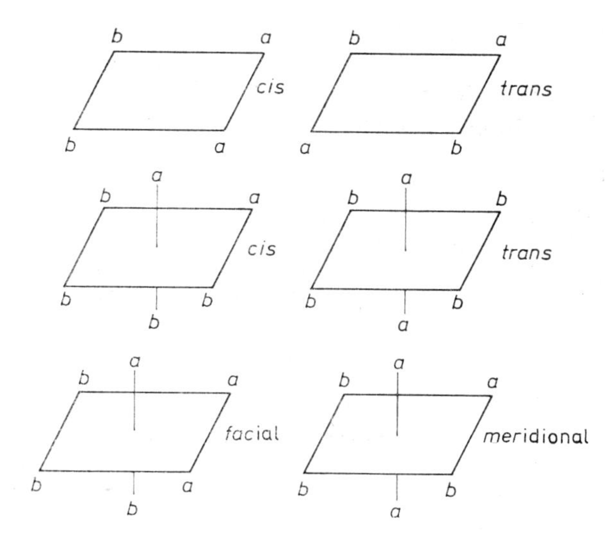

For representing structures of coordination compounds it is usually simpler and clearer to use the geometrical solid whose vertices represent the positions of the coordinated ligands: *e.g.*, the square (above) and octahedron (**7.511**, Examples 5–8) for 4-planar and 6-octahedral coordination. Also, for 6-octahedral coordination the mixed representation (plane and axis) (above) is frequently used. Often the central atom is omitted in such representations.

(e) Chiral (asymmetrical) arrangement of ligands in the coordination sphere (see **7.8**).

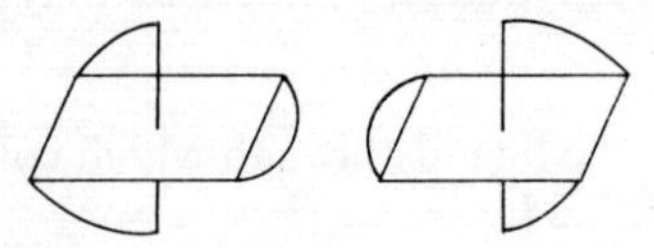

∪ represents a bidentate group such as $H_2NCH_2CH_2NH_2$ ("en")

(f) Asymmetry of an atom in a ligand which originates in the coordination process: for example, in the coordination of the bidentate ligand below, the carbon marked with an asterisk is rendered asymmetric.

Cl_4Pt — NH_2—CH_2 / NH_2—*CH—CH_2NH_2

7.51—Geometrical Isomerism

7.511—The prefixes *cis*-, *trans*-, *fac*-, and *mer*- are used where they are sufficient to designate specific isomers (Table III).

Examples:

Planar configuration

1. Et_3Sb, Et_3Sb, Pt, I, I — *cis*-diiodobis(triethylstibine)platinum
cis-diiodobis(triethylstibine)platinum(II)

2. Pr_2S, NO_2, Pt, O_2N, SPr_2 — *trans*-bis(dipropyl sulfide)dinitroplatinum
trans-bis(dipropyl sulfide)dinitroplatinum(II)

3. 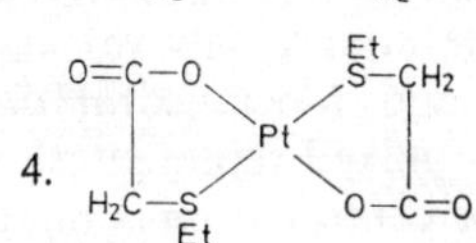
cis-bis(glycinato-*O*,*N*)platinum
cis-bis(glycinato-*O*,*N*)platinum(II)

4. O=C—O, Pt, H_2C—S(Et), Et-S—CH_2, O—C=O — *trans*-bis[(ethylthio)acetato-*O*,*S*]platinum
trans-bis[(ethylthio)acetato-*O*,*S*]platinum(II)

Octahedral configuration*

5. *cis*-bis(ethylenediamine)difluorocobalt(1+) ion
cis-bis(ethylenediamine)difluorocobalt(III) ion

6. *trans*-tetraamminedichlorochromium(1+) ion
trans-tetraamminedichlorochromium(III) ion

7. *fac*-trichlorotris(pyridine)ruthenium
fac-trichlorotris(pyridine)ruthenium(III)

8. 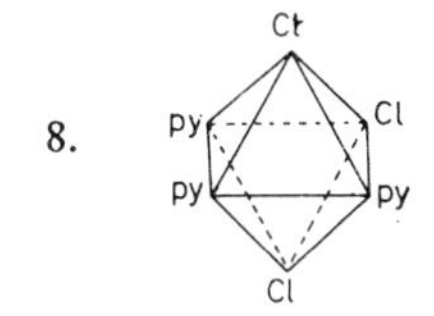

mer-trichlorotris(pyridine)ruthenium
mer-trichlorotris(pyridine)ruthenium(III)

7.512—Italicized letters are used as locant designators for specifying spatial positions in various configurations. The assignments for the square planar and octahedral configurations are:

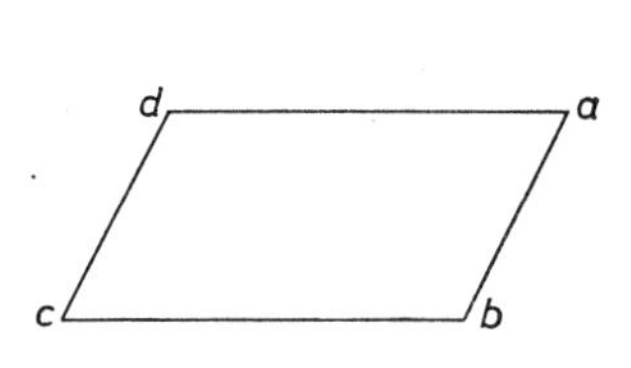

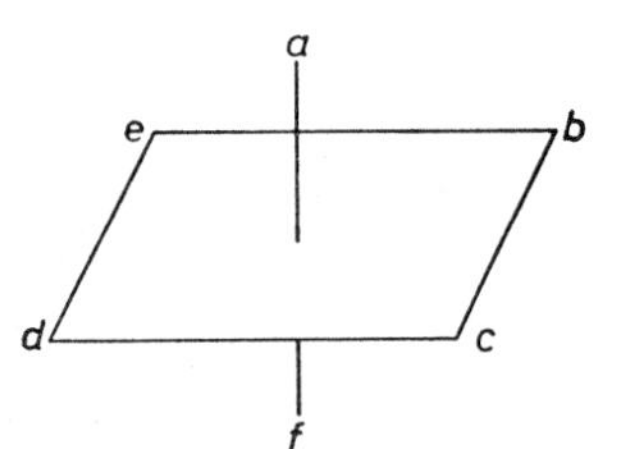

In the past, numerals have been used as locant designators in the nomenclature of coordination entities. However, with the many uses of numbers for other purposes in accepted nomenclature patterns, it seems best to use small letters as locant designators within the coordination sphere. Further, in giving a series of letters, it is not necessary to place commas between each letter as would be required with numbers. The use of letter locants has further merit because there is an essential difference between the use of locants for designating positions in space around a central atom and for designating substitution on a particular atom of a group of atoms.

* In these and in some of the following examples the central metal atom has been omitted from the formulae for the sake of clarity.

The first ligand to be mentioned in the name is given the lowest possible designator and the second ligand, the next lowest possible designator. The assignments to the remaining ligands follow from their position in the complex as lettered above.

By the choice of a specific order and direction for assigning locants, it is possible to designate a specific optical isomer and to distinguish between enantiomers. However, in contrast to the practice for organic enantiomers (with tetrahedral atoms) there are differences in the names for the enantiomers other than some symbol designating the chirality. There is objection to this practice by some who prefer that the assignment of locants for one optically active form (say the levo) be the mirror image of that for the dextro. This matter of the choice of a system of assignment of locants for enantiomers is under careful consideration by the Commission.

Examples:

1.

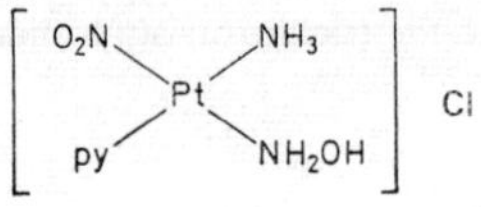

O_2N NH_3
Pt Cl
py NH_2OH

a-ammine-*b*-(hydroxylamine)-*d*-nitro-*c*-(pyridine)platinum(1+) chloride
a-ammine-*b*-(hydroxylamine)-*d*-nitro-*c*-(pyridine)platinum(II) chloride

There are two more isomers of this composition.

2.

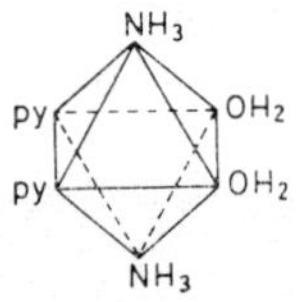

NH_3
py OH_2
py OH_2
NH_3

af-diammine-*bc*-diaqua-*de*-bis(pyridine)cobalt(3+) ion
af-diammine-*bc*-diaqua-*de*-bis(pyridine)cobalt(III) ion

There are four more possible isomers of this composition, one of which exists in enantiomeric forms.

7.513—The application of these locant designators to chelate ligands is based on the following rules*:

(a) For symmetrical linear ligands of the type A————A, the position in the coordination sphere of the coordinating atom at one end of the ligand shall be given and then, successively, that of each atom of the ligand through which coordination takes place.

* These rules are applicable to ligands which are strictly linear and those in which the linear portion is also part of another ring. The atoms of the ring other than the linear portion may be considered as substituents.

Examples:

1.

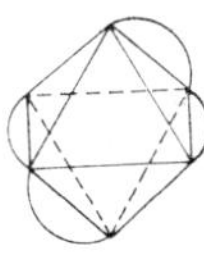

◡◡ represents $H_2NCH_2CH(NH_2)CH_2NH_2$

abc,edf-bis(1,2,3-propanetriamine)cobalt(3+) ion
abc,edf-bis(1,2,3-propanetriamine)cobalt(III) on

There are two more isomers of this composition, one of which exists in enantiomeric forms.

2.

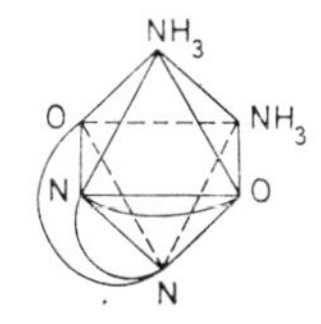

O◡N◡N◡O represents $\{CH_3C(O^-){=}CH{-}C(CH_3){=}NCH_2{-}\}_2$

ab-diammine-*cdfe*-[*N,N′*-ethylenebis(4-imino-2-pentanonato)(2−)-*O,N,N′,O′*]cobalt(1+) ion
ab-diammine-*cdfe*-[*N,N′*-ethylenebis(4-imino-2-pentanonato)(2−)-*O,N,N′,O′*]cobalt(III) ion

Of the three isomers of this composition, two may exist in enantiomeric forms.

(b) For unsymmetrical linear ligands of the type A————X, the positions of each coordinating atom in the ligand in the coordination sphere shall be given successively starting at one end. The coordinating atom to be cited first shall be chosen on the basis of the following preferences whichever first applies.

(i) Difference in end groups which can coordinate: the end with the element which occurs earlier in the series in Table IV.

Examples:

1. $H_2NCH_2CH_2SR$ —S first
2. $^-OOCCH_2NH_2$ —O first
3. *o*-$R_2AsC_6H_4PR_2$ —P first

(ii) Difference in substitution at ends beyond the coordination sites or between them: the site with the lowest number according to organic practice of naming carbon compounds.

Examples:

4. $CF_3COCHCOCH_3^-$
1,1,1,-trifluoro-2,4-pentanedionato, CF_3CO—first
5. $H_2NCH(CH_3)CH_2NH_2$
1,2-propanediamine, H_2NCH_2—first

(iii) Difference in ends where the *same element* is in a different linkage: the end which is listed first in the following table (*cf. IUPAC Nomenclature of Organic Chemistry*, 1971, p. 87, **C–10.4**). An atom in a chain has priority over an atom of the same element in a ring.

O: —COO^-, —CHO, C=O, —OH, —O^-, —OR
S: —CSS^-, —CHS, C=S, —SH, —S^-, —SR
N: —$CONH_2$, —CN, —CH=N—, C=N—, —NH_2,
—NH^-, —NHR, —NR^-, —NR_2, hydrazines

Examples:

6. O^- / CHO —CHO first

7. $CH_3NHCH_2CH_2NH_2$ —NH_2 first

8. N / CH_2NH_2 —NH_2 first

9. O / CH_2OCH_3 —OCH_3 first

(iv) Same end groups but difference in the penultimate (or antepenultimate) sites: the end nearest the element which occurs earlier in Table IV (*cf.* i above).

Examples:

10. $H_2NCH_2CH_2SCH_2CH_2OCH_2CH_2NH_2$ —NH_2 nearest O first
11. $^-OOCCH_2NRCH_2CH_2SCH_2COO^-$ —COO^- nearest S first
12. *o*-$^-OC_6H_4CH{=}NCH_2CH_2AsRCH_2CH_2SCH_2CH_2N{=}CHC_6H_4O^-$(*o*) —$O^-$ nearest S first
13. $^-OOCCH_2NRCH_2PO(OH)O^-$ O attached to P first
14. $^-O(HO)BCH_2NHCH_2SO_2O^-$ O attached to S first

(v) Same end coordination site attached to identical atoms: the end attached to the atom in lower oxidation state.

Examples:

15. $^-O(O)HPCH_2NRCH_2PO(OH)O^-$ $^-O(O)HP$— first
16. $^-O(RO)PCH_2NRCH_2PO(OR)O^-$ $^-O(RO)P$— first

(vi) Same end groups but differing numbers of carbon atoms to next coordinating site: the end forming the smaller chelate ring.

Examples:

17. $H_2\overset{*}{N}CH_2CH_2OCH_2CH_2CH_2NH_2$ N marked * first
18. $^-\overset{*}{O}OCCH_2NHCH_2CH_2COO^-$ O marked * first
19.

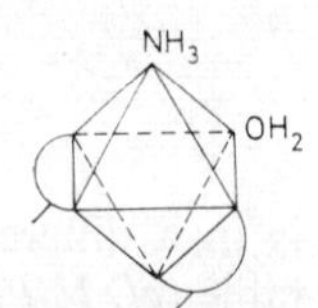

⌣ represents $H_2NCH_2CH(NH_2)CH_3$
a-ammine-*b*-aqua-*cf*,*ed*-bis(1,2-propanediamine)cobalt(3+) ion
a-ammine-*b*-aqua-*cf*,*ed*-bis(1,2-propanediamine)cobalt(III) ion

There are six isomers of this composition, four of which exist as enantiomeric pairs.

20.

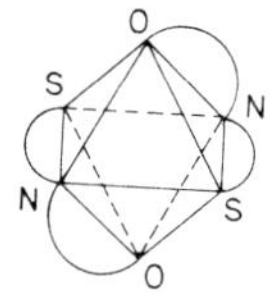

S◡N◡O represents $CH_3SCH_2CH_2N{=}CHC_6H_4O^{-}(o)$

abc,fde-bis{*o*-{*N*-[2-(methylthio)ethyl]formimidoyl}-phenolato-*O,N,S*}cobalt(1+) ion

abc,fde-bis{*o*-{*N*-[2-(methylthio)ethyl]formimidoyl}-phenolato-*O,N,S*}cobalt(III) ion

Six enantiomeric pairs of isomers of this composition are expected.

(c) Symmetrical branched ligands

(i) For symmetrical branched ligands of the types A_2———A_2, the position in the coordination sphere of either of the coordinating atoms at one end of the ligand shall be given, then the position of the corresponding atom in the branch, followed by those of the intermediate coordinating atoms and completed by the positions of the two terminal coordinating atoms.

Example:

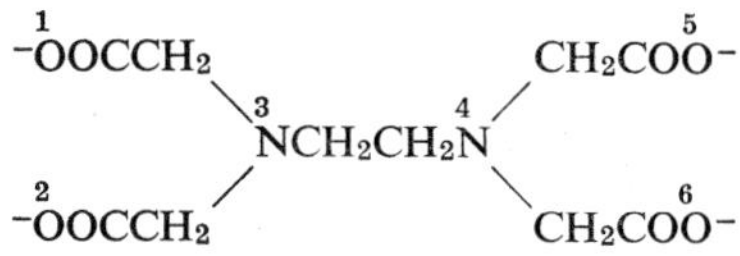

```
  1                               5
-OOCCH2                     CH2COO-
       \3              4  /
        NCH2CH2N
  2    /                  \       6
-OOCCH2                     CH2COO-
```

(ii) For symmetrical branched ligands of the type $(AB)_2$———$(BA)_2$ the order shall be the same as for the type A_2———A_2 except that the positions of all coordinating atoms on one branch shall be given before proceeding to those in the corresponding branch.

Example:

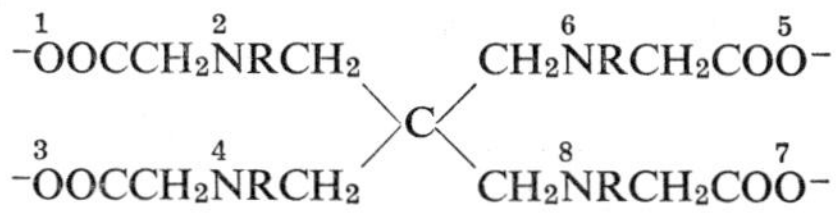

```
  1        2               6         5
-OOCCH2NRCH2          CH2NRCH2COO-
                \C/
  3        4   /   \       8         7
-OOCCH2NRCH2          CH2NRCH2COO-
```

(d) For unsymmetrical branched ligands the rules under (b) are used to determine which end shall be designated first. The listing of the other sites shall follow as in (c).

Examples:

1. $(^{-}OOCCH_2)_2NCH_2CH_2SCH_3$ — the O's first
2. $(^{-}OOCCH_2)_2NCH_2CH_2SCH_2COO^{-}$ — the O nearest S first
3. *o*-$^{-}OC_6H_4CH(COO^{-})NHCH_2CH_2NH_2$ — the O of —COO^{-} first
4. $(^{-}OOCCH_2)_2NCH_2CH(CH_3)\overset{*}{N}(CH_2COO^{-})_2$ — the O's nearest N marked * first

5.

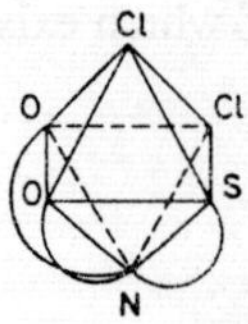

$S\smile N(\smile O)_2$ represents $CH_3\overset{4}{S}CH_2CH_2\overset{3}{N}(\overset{1}{CH_2COO^-})(\overset{2}{CH_2COO^-})$

ab-dichloro-*defc*-{[2-(methylthio)ethyl]iminodiacetato-*O*,*O*′,*N*,*S*}platinum
ab-dichloro-*defc*-{[2-(methylthio)ethyl]iminodiacetato-*O*,*O*′,*N*,*S*}platinum(IV)

There is an enantiomer of the above compound and another inactive isomer.

6.

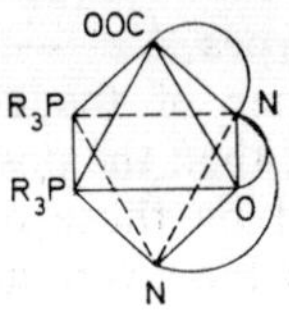

$N\frown N\smile COO$ (with $O\smile$) represents $H_2\overset{4}{N}CH_2CH_2\overset{3}{N}HCH[C_6H_4\overset{2}{O}{}^-(o)]CO\overset{1}{O}{}^-$

abcf-[*N*-(2-aminoethyl)-2-(*o*-hydroxyphenyl)glycinato(2−)-*O*,*O*′,*N*,*N*′]-*de*-bis(tributylphosphine)cobalt(1+) ion
abcf-[*N*-(2-aminoethyl)-2-(*o*-hydroxyphenyl)glycinato(2−)-*O*,*O*′,*N*,*N*′]-*de*-bis(tributylphosphine)cobalt(III) ion

There are three enantiomeric pairs of this composition.

(e) For ligands with central branching of the types A—(K)—A and A—(N)—X, the sites on the linear portion of the ligand shall be designated in the normal manner with that of the central branch in parentheses interposed at (or nearest to) the point of attachment.

This rule concerns only sexidentate ligands and those of higher function because central branching in ligands of lower functionality reduces to cases covered previously: A(B)C; A(B)CD, *etc.*

Examples of numbering:

1. $^-\overset{1}{O}OCCH_2\overset{2}{N}(CH_3)CH_2CH_2\overset{3}{N}(\overset{4}{CH_2COO^-})CH_2CH_2\overset{5}{N}(CH_3)CH_2CO\overset{6}{O}{}^-$

2. $H_2\overset{6}{N}CH_2CH_2\overset{5}{N}HCH_2CH_2\overset{3}{N}(CH_2CH_2\overset{4}{N}H_2)CH_2CH_2\overset{2}{S}CH_2CH_2\overset{1}{N}H_2$

Example:

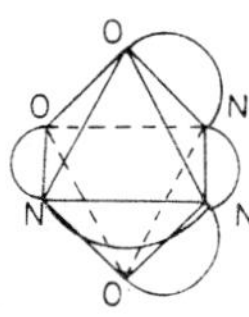

O◡N(◡N◡O)$_2$ represents $^{-}OOCCH_2N\{CH_2CH_2N(CH_3)CH_2COO^{-}\}_2$

abc(*f*)*de*-[bis(2-methylaminoethyl)amine-*N*,*N*′,*N*″-triacetato-*O*,*N*,*N*′,*O*′,*N*″,*O*″]cobalt

abc(*f*)*de*-[bis(2-methylaminoethyl)amine-*N*,*N*′,*N*″-triacetato-*O*,*N*,*N*′,*O*′,*N*″,*O*″]cobalt(III)

There are three enantiomeric pairs of this composition.

7.514—The assignment of locant designators for other configurations around a coordination centre is based on locating planes of atoms perpendicular to a major axis in each configuration and assigning locants in a fixed manner in each successive plane. The actual procedure is: first, locate the highest (and longest in case of a choice) order axis of rotational symmetry; second, where the axis is not symmetrical, choose that end with a single atom (or smallest number of atoms) in the first plane to be numbered; third, locate the first plane of atoms (atom) to receive locants; fourth, orient the molecule so that the first position to receive a locant in the first plane with more than one atom is in the twelve o'clock position; fifth, assign locant designators to the axial position or to each coordinating position in the first plane, beginning at the 12 o'clock position and moving in a clockwise direction; sixth, from the first plane, move to the next position and continue assignments in the same manner, always returning to the 12 o'clock position or in the position nearest to it clockwise before assigning any locants in that plane; seventh, continue this operation until all positions are assigned.*

Figures illustrating this procedure for coordination numbers 5 and 8 together with their McDONNELL–PASTERNAK class symbols [*J. Chem. Document.*, **5**, 57 (1965)] are given below. Other less common configurations may similarly be designated by their class symbols.

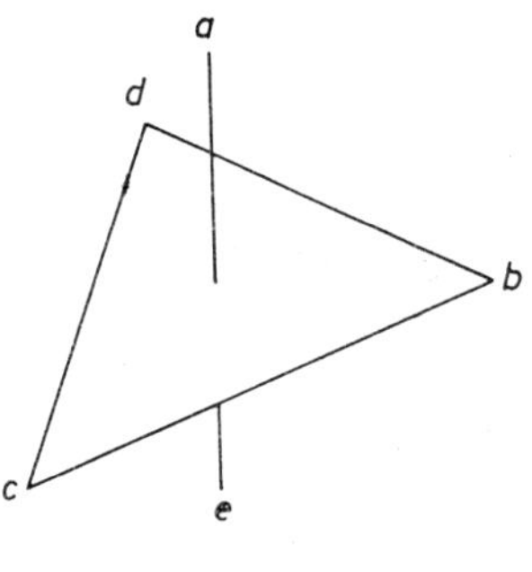

trigonal bipyramid
5A

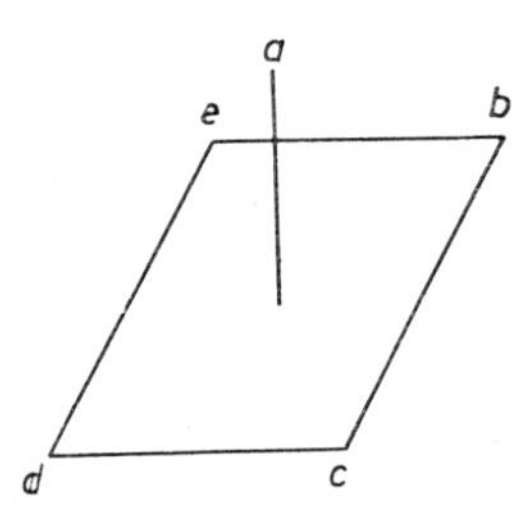

square pyramid
5B

* This procedure is consistent with that used for the assignment of locants in boron hydrides, *I.U.P.A.C. Information Bulletin:* Appendices on Tentative Nomenclature, Symbols, Units and Standards, No. 8 (September 1970) or *Inorg. Chem.*, **7**, 1948 (1968).

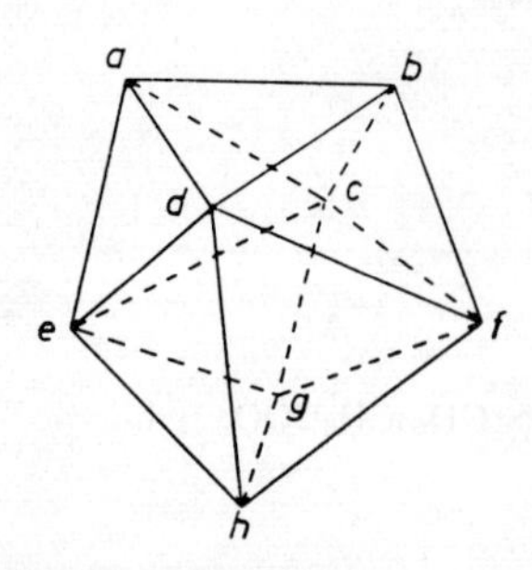

digonal dodecahedron
8A

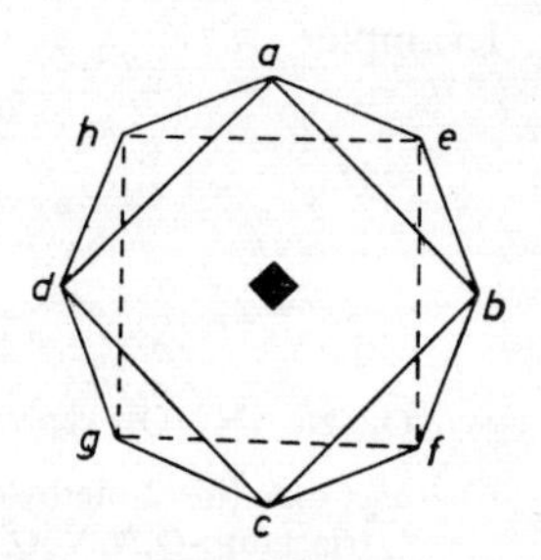

square antiprism
8B

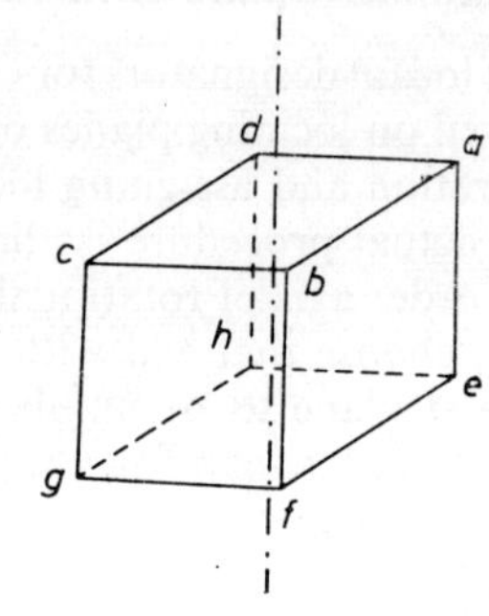

cube (regular hexahedron)
8D

Although chemists have become accustomed to visualizing the arrangement in space of an octahedron, *etc.*, from two-dimensional drawings, there is greater difficulty visualizing a dodecahedron, *etc.* A convenient device for visualizing the assignment of locant designators is the representation of the successive planes and the atoms contained in the planes. Thus, the dodecahedron 8A is seen to involve the following unit planes of atoms:

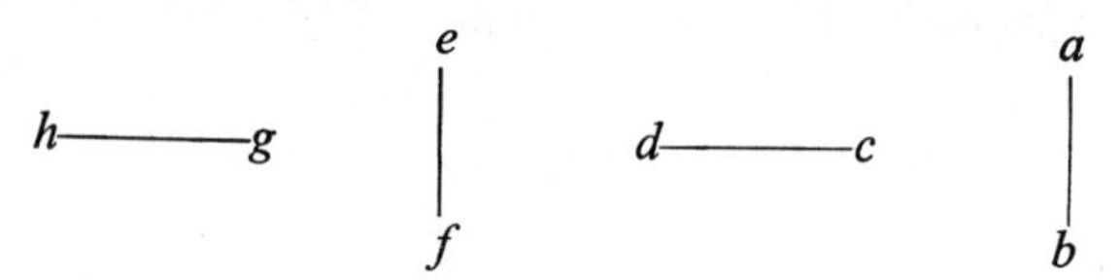

7.52—Isomerism due to Chirality (Asymmetry); see also 7.8

Locant designators distinguish enantiomeric forms of a coordination compound.

Examples:

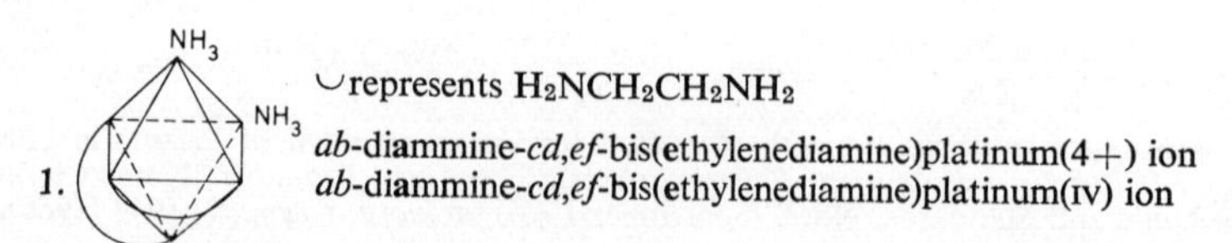

1. ◡ represents $H_2NCH_2CH_2NH_2$

ab-diammine-*cd,ef*-bis(ethylenediamine)platinum(4+) ion
ab-diammine-*cd,ef*-bis(ethylenediamine)platinum(IV) ion

2. 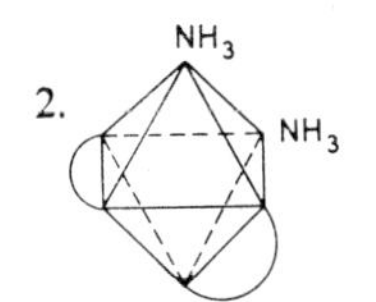

◡ represents $H_2NCH_2CH_2NH_2$

ab-diammine-*cf,de*-bis(ethylenediamine)platinum(4+) ion
ab-diammine-*cf,de*-bis(ethylenediamine)platinum(IV) ion

This is an enantiomer of the preceding example.

3. 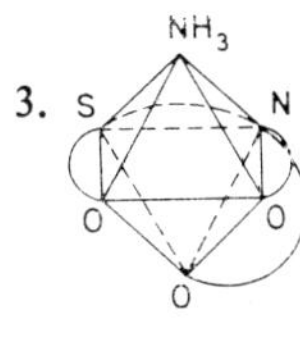

O◡S◡N(◡O)₂ represents $^{-}\overset{1}{O}OCCH_2\overset{2}{S}CH_2CH_2\overset{3}{N}(\overset{4}{C}H_2COO^{-})(\overset{5}{C}H_2COO^{-})$

a-ammine-*bcdef*-[2-(carboxymethylthio)ethyliminodiacetato(3−)-*O,S,N,O′,O″*]chromium
a-ammine-*bcdef*-[2-(carboxymethylthio)ethyliminodiacetato(3−)-*O,S,N,O′,O″*]chromium(III)

4.

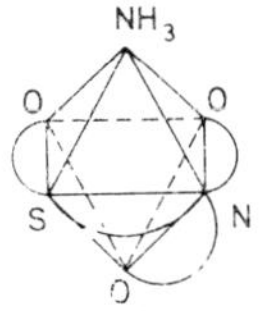

a-ammine-*bedcf*-[2-(carboxymethylthio)ethyliminodiacetato(3−)-*O,S,N,O′,O″*]chromium
a-ammine-*bedcf*-[2-(carboxymethylthio)ethyliminodiacetato(3−)-*O,S,N,O′,O″*]chromium(III)

This is an enantiomer of the preceding example.

When the absolute configuration of a coordination compound is not known, locant designators may still be used, but the complete name should be prefixed '*X*'. The observed sign of rotation + or − may also be at any specified wavelength, *e.g.*

$(+)_{589}$*X*-*a*-ammine-*bcdef*-[2-(carboxymethylthio)ethyliminodiacetato(3−)-*O,S,N,O′,O″*]chromium

is the name of example 3 or its enantiomer example 4 whichever is dextrorotatory at 589 nm. When the substance is known to be a mixture of racemic forms, the name of one enantiomeric form should be prefixed '*rac*', *e.g.* the name

rac-*ab*-diammine-*cd,ef*-bis(ethylenediamine)platinum(4+) ion

describes a racemic mixture of examples 1 and 2 above.

7.6. DI- AND POLYNUCLEAR COMPOUNDS WITH BRIDGING GROUPS

7.61—Compounds with Bridging Atoms or Groups

7.611—(a) A bridging group is indicated by adding the Greek letter μ immediately before its name and separating the name from the rest of the complex by hyphens.

(b) Two or more bridging groups of the same kind are indicated by di-μ- (or bis-μ-), *etc.*

(c) The bridging groups are listed with the other groups in alphabetical order *unless the symmetry of the molecule permits simpler names by the use of multiplicative prefixes* (*cf.* Example 1).

(d) Where the same ligand is present as a bridging ligand and as a non-bridging ligand, it is cited first as a bridging ligand.

Bridging groups between two centres of coordination are of two types: (1) the two centres are attached to the same atom of the bridging group and (2) the two centres are attached to different atoms of the bridging group. For bridging groups of the first type it is often desirable to indicate the bridging atom. This is done by adding the italicized symbol for the atom at the end of the name of the ligand as in **7.33**. For bridging groups of the second type, the symbols of all coordinated atoms are added.

Examples:

1. $[(NH_3)_5Cr—OH—Cr(NH_3)_5]\,Cl_5$
   μ-hydroxo-bis(pentaamminechromium)(5+) chloride
   μ-hydroxo-bis[pentaaminechromium(III)] chloride
2. $[(CO)_3Fe(CO)_3Fe(CO)_3]$
   tri-μ-carbonyl-bis(tricarbonyliron)
3. $[Br_2Pt(SMe_2)_2PtBr_2]$
   bis(μ-dimethyl sulfide)-bis[dibromoplatinum(II)]
4. $[(CO)_2Ni(Me_2PCH_2CH_2PMe_2)_2Ni(CO)_2]$
   bis[μ-ethylenebis(dimethylphosphine)]-bis(dicarbonylnickel)
5. $[(CO)\{P(OEt)_3\}Co(CO)_2Co(CO)\{P(OEt)_3\}]$
   di-μ-carbonyl-bis[carbonyl(triethyl phosphite)cobalt]
6. $[\{Au(CN)(C_3H_7)_2\}_4]$
   cyclo-tetra-μ-cyano-tetrakis(dipropylgold) (for use of *cyclo* see Table III)
7. $[\{(MoF_4)F\}_4]$
   cyclo-tetra-μ-fluoro-tetrakis(tetrafluoromolybdenum)
   cyclo-tetra-μ-fluoro-tetrakis[tetrafluoromolybdenum(V)]
8. 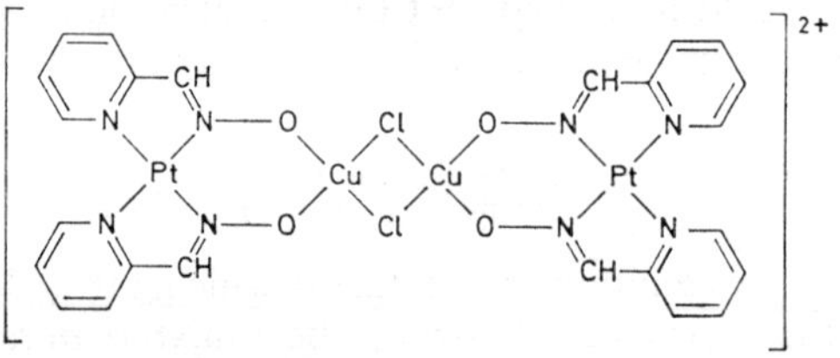 di-μ-chloro-bis{[bis(picolinaldehyde oximato-*N,N′*)platinum-*O,O′*] copper}(2+) ion
   di-μ-chloro-bis{[bis(picolinaldehyde oximato-*N,N′*)platinum(II)-*O,O′*] copper(II)} ion
9. 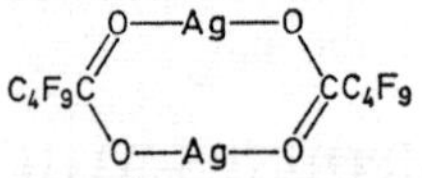 bis(μ-nonafluorovalerato-*O,O′*)-disilver
   bis(μ-nonafluorovalerato-*O,O′*)-disilver(I)
10. $(ON)_2Fe(SC_2H_5)_2Fe(NO)_2$
    bis(μ-ethylthio)-tetranitrosyldiiron
11. 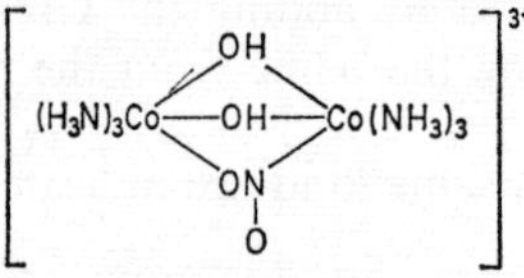 hexaammine-di-μ-hydroxo-μ-nitrito(*O,N*)-dicobalt(3+) ion
    hexaammine-di-μ-hydroxo-μ-nitrito(*O,N*)-dicobalt(III) ion
12. 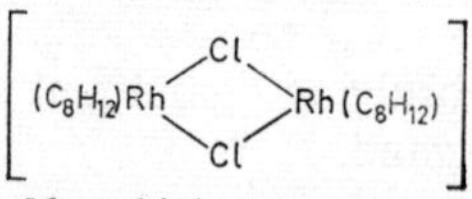 di-μ-chloro-bis(η-1,5-cyclooctadiene)dirhodium
    di-μ-chloro-bis(η-1,5-cyclooctadiene)dirhodium(I)

7.612—If the number of central atoms bound by one bridging group exceeds two, the number shall be indicated by adding a subscript numeral to the μ.

Examples:

1. $[\{PtI(CH_3)_3\}_4]$
   tetra-μ_3-iodo-tetrakis(trimethylplatinum)
   tetra-μ_3-iodo-tetrakis[trimethylplatinum(IV)]
2. $[Be_4O(CH_3COO)_6]$
   hexa-μ-acetato-(*O*,*O*′)-μ_4-oxo-tetraberyllium
   hexa-μ-acetato-(*O*,*O*′)-μ_4-oxo-tetraberyllium(II)
3. $[Cr_3O(CH_3COO)_6]$ Cl
   hexa-μ-acetato-(*O*,*O*′)-μ_3-oxo-trichromium(1+) chloride
   hexa-μ-acetato-(*O*,*O*′)-μ_3-oxo-trichromium(III) chloride
4. $[(CH_3Hg)_4S]^{2+}$
   μ_4-thio-tetrakis(methylmercury)(2+) ion
   μ_4-thio-tetrakis[methylmercury(II)] ion

7.613—More complicated structures are named by the use of locant designators. The principles of **7.514** for assigning locant designators are used with the stipulation that the "axis" of operation must be chosen so as to pass continuously through the largest number of the nuclear atoms even if this results in a bent axis. The following assignments result:

(The planes of atoms are shown beside each structure).

1.
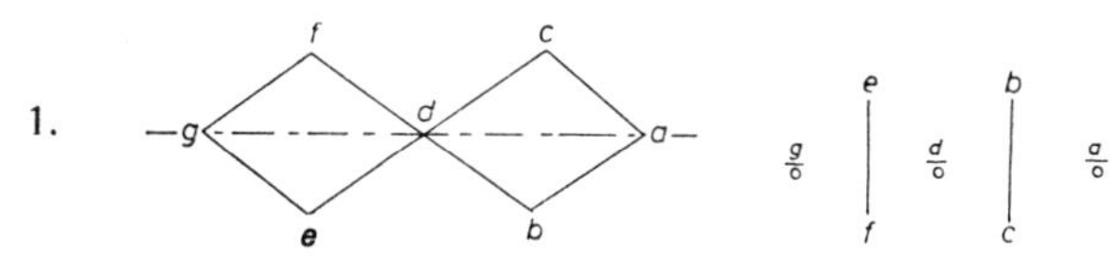

2.
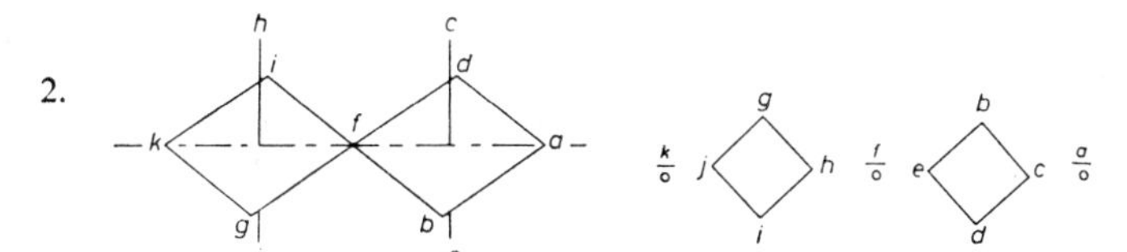

3.
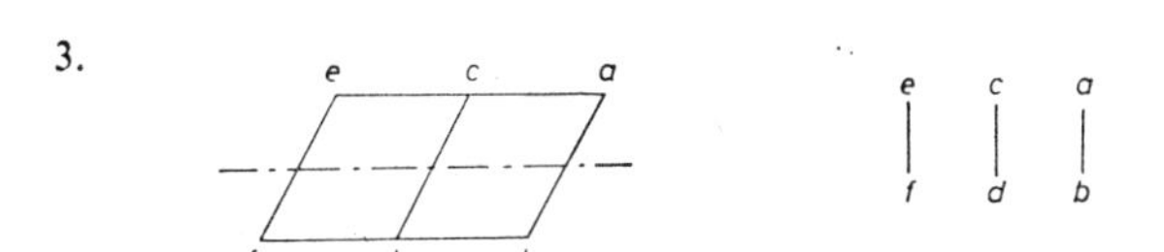

4.
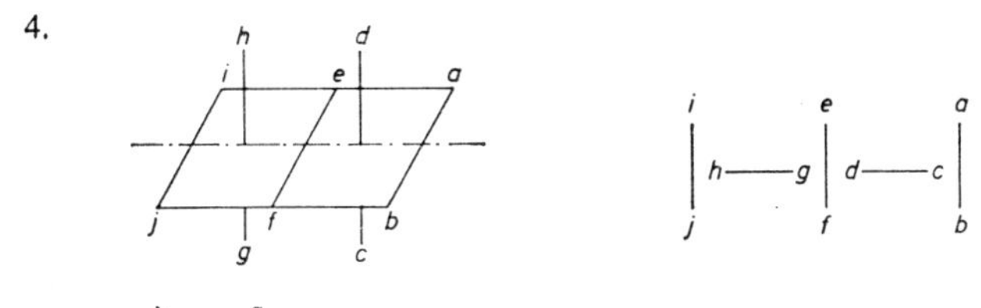

5.
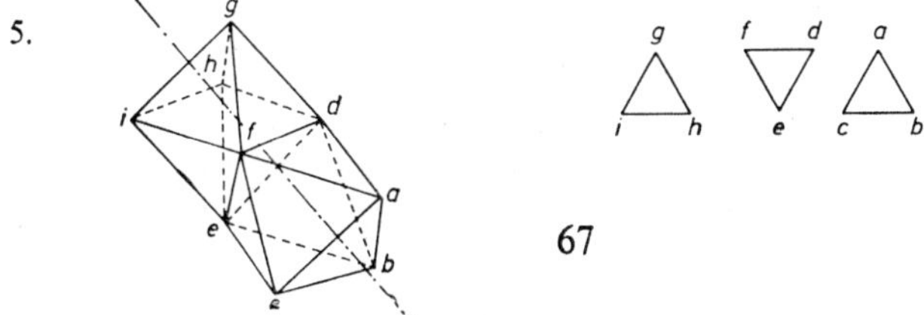

6.

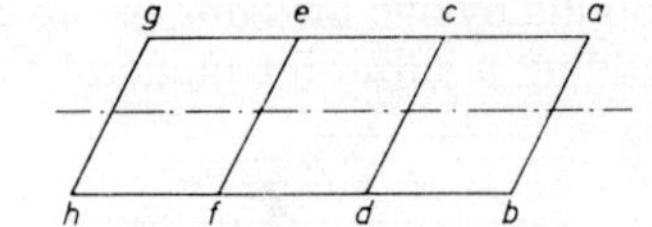

7.

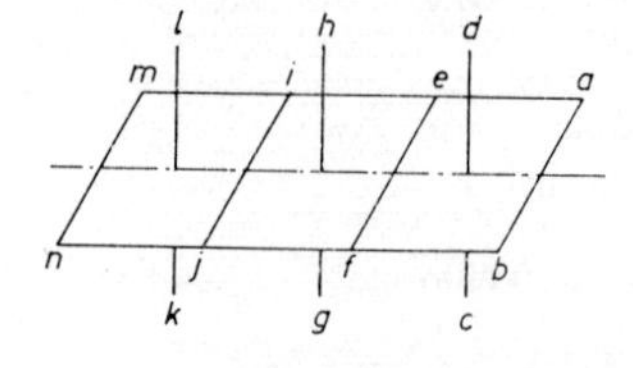

8.

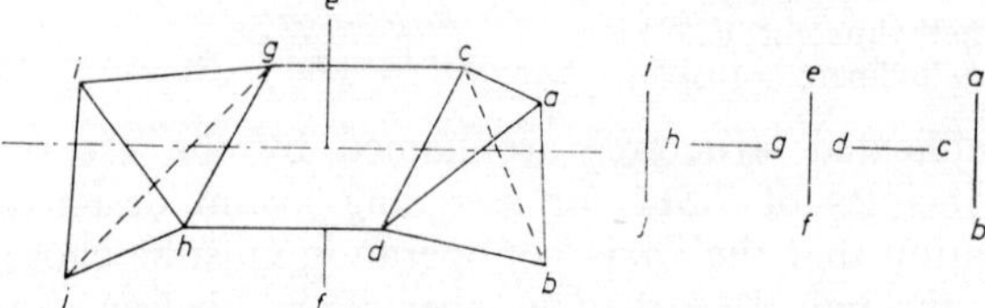

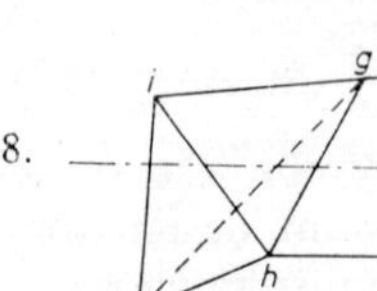

9.

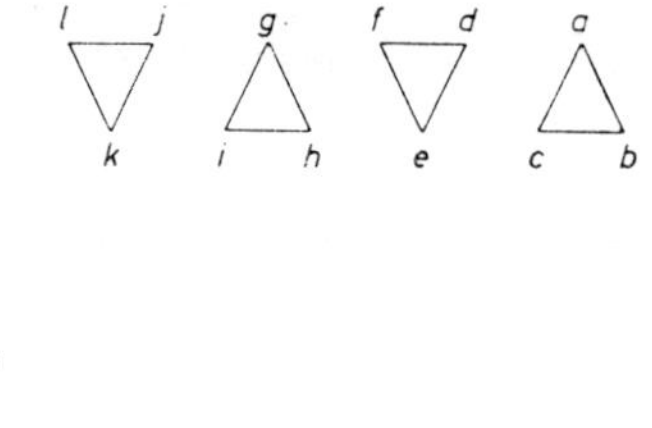

Examples:

1.

af-bis(**b**utylamine)-di-μ-**c**hloro-di**c**hlorodiplatinum
af-bis(**b**utylamine)-di-μ-**c**hloro-di**c**hlorodiplatinum(II)

2.

di-μ-**c**hloro-*ae*-di**c**hlorobis(**t**riethylarsine)diplatinum
di-μ-**c**hloro-*ae*-di**c**hlorobis(**t**riethylarsine)diplatinum(II)

3.

di-μ-**c**hloro-*ab*-di**c**hlorobis(**t**riethylarsine)diplatinum (unknown)
di-μ-**c**hloro-*ab*-di**c**hlorobis(**t**riethylarsine)diplatinum(II) (unknown)

4. [NCS, SCN, $P(C_3H_7)_3$ / Pt Pt / $(C_3H_7)_3P$ NCS SCN]

di-μ-thiocyanato-*S*,*N*-*af*-dithiocyanatobis(tripropylphosphine)diplatinum
di-μ-thiocyanato-*S*,*N*-*af*-dithiocyanatobis(tripropylphosphine)diplatinum(II)

5. EtS S Br / Pd Pd / Br S SEt

EtS◡S represents $C_2H_5SCH_2CH_2S^-$
af-dibromo-*bd*,*ec*-bis[μ-(2-ethylthio-*S*′)-ethylthio-*S*′,μ-*S*]-dipalladium
af-dibromo-*bd*,*ec*-bis[μ-(2-ethylthio-*S*′)-ethylthio-*S*′,μ-*S*]-dipalladium(II)

6.

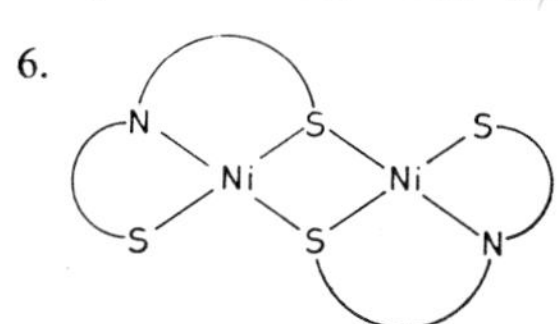

S◡N◡S represents $CH_3N(CH_2CH_2S^-)_2$
abd,*cef*-bis-{μ-[2,2′-(methylimino)bis(ethylthio)(2−)]-*S*,*N*,μ-*S*′}dinickel
abd,*cef*-bis-{μ-[2,2′-(methylimino)bis(ethylthio)(2−)]-*S*,*N*,μ-*S*′}dinickel(II)

7.

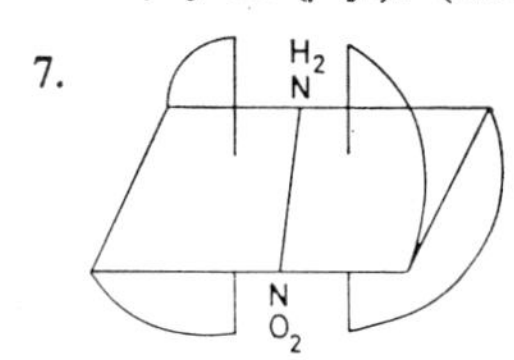

◡represents $H_2NCH_2CH_2NH_2$
e-μ-amido-*ac*,*bd*,*gj*,*hi*-tetrakis(ethylenediamine)-*f*-μ-nitrodicobalt(4+) ion
e-μ-amido-*ac*,*bd*,*gj*,*hi*-tetrakis(ethylenediamine)-*f*-μ-nitrodicobalt(III) ion
The enantiomer is:
e-μ-amido-*ad*,*bc*,*gi*,*hj*-tetrakis(ethylenediamine)-*f*-μ-nitrodicobalt(4+) ion
e-μ-amido-*ad*,*bc*,*gi*,*hj*-tetrakis(ethylenediamine)-*f*-μ-nitrodicobalt(III) ion
The meso form is:
e-μ-amido-*ad*,*bc*,*gj*,*hi*-tetrakis(ethylenediamine)-*f*-μ-nitrodicobalt(4+) ion
e-μ-amido-*ad*,*bc*,*gj*,*hi*-tetrakis(ethylenediamine)-*f*-μ-nitrodicobalt(III) ion

8.

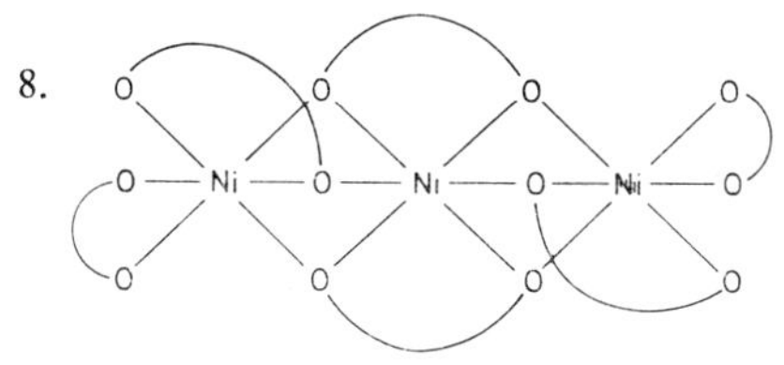

O◡O represents $(CH_3COCHCOCH_3)^-$
bis(2,4-pentanedionato)nickel trimer or
ad,*ik*-bis[μ-(2,4-pentanedionato)-μ-*O*,*O*′]-*eh*,*fg*-bis[μ_3-(2,4-pentanedionato)-μ-*O*,μ-*O*′]-*bc*,*jl*-bis(2,4-pentanedionato)trinickel
bis(2,4-pentanedionato)nickel(II) trimer or
ad,*ik*-bis[μ-(2,4-pentanedionato)-μ-*O*,*O*′]-*eh*,*fg*-bis[μ_3-(2,4-pentanedionato)-μ-*O*,μ-*O*′]-*bc*,*jl*-bis(2,4-pentanedionato)trinickel(II)

9.

abcdgijlopqr-dodecaammine-*efhkmn*-hexa-μ-hydroxo-tetrachromium(6+) ion
abcdgijlopqr-dodecaammine-*efhkmn*-hexa-μ-hydroxo-tetrachromium(III) ion

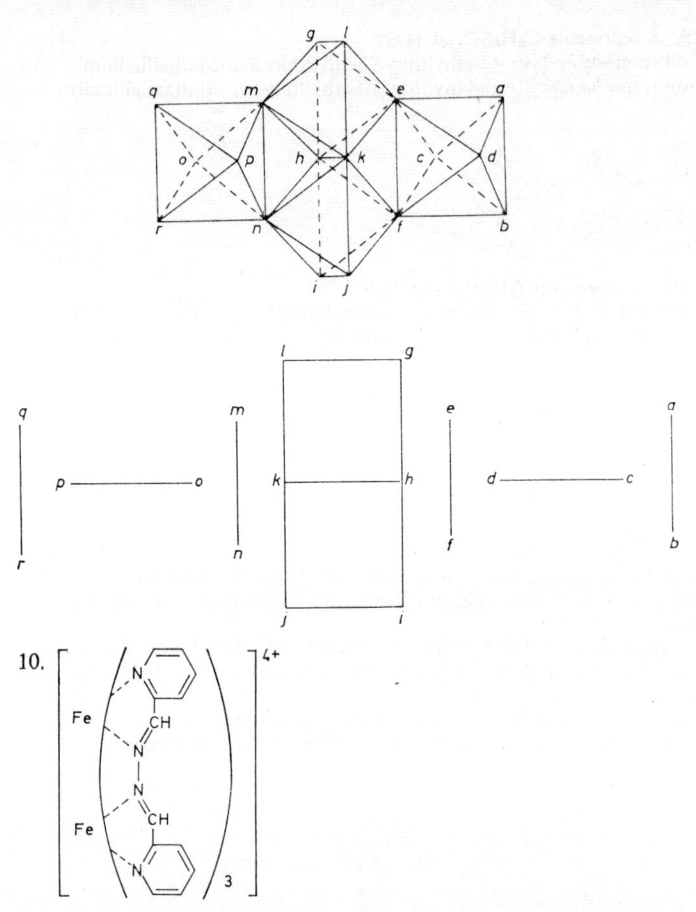

adhk,beil,cfgj-tris[μ-(picolinaldehyde azine)-*N,N′,N″,N‴*]-diiron(4+) ion
adhk,beil,cfgj-tris[μ-(picolinaldehyde azine)-*N,N′,N″,N‴*]-diiron(II) ion

If the locant indicators are assigned to each of the two iron atoms independently no simplification of the name is attained but a new principle is involved: *adb′e′*, *bec′f′*, *cfa′d′*-tris[μ-(picolinaldehyde azine)-*N,N′,N″,N‴*]-diiron(4+) or *adb′e′*, *bec′f′*, *cfa′d′*-tris[μ-(picolinaldehyde azine)-*N,N′,N″,N‴*]-diiron(II) ion.

7.614—When the bridged polynuclear complex contains more than one kind of nuclear atom, locant designators shall be assigned in such a manner that the nuclear atom coming first in Table IV shall receive the site with locant *a*. When the end nuclear atoms are the same, the end with locant *a* is similarly determined by the penultimate nuclear atom. When the order of nuclear atoms is symmetrical the alphabetical order of ligands determines the assignment of locant *a*.

The nuclear atoms shall be listed in the order in which they are present in the compound starting with the atom having locant *a*.

Example:

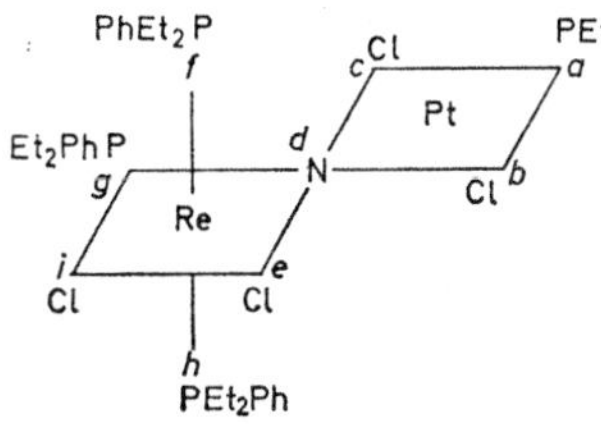

$=[Cl_2(PEt_2Ph)_3Re{\equiv}N{-}PtCl_2(PEt_3)]$

bcei-tetrachloro-*fgh*-tris(diethylphenylphosphine)-*d*-μ-nitrido-*a*-(triethylphosphine)platinumrhenium

bcei-tetrachloro-*fgh*-tris(diethylphenylphosphine)-*d*-μ-nitrido-*a*-(triethylphosphine)platinum(II)rhenium(V)

7.62—Extended Structures

Where bridging causes an indefinite extension of the structure it is best to name compounds primarily on the basis of the repeating unit; thus the compound having the composition represented by the formula $CsCuCl_3$ has an anion with the structure:

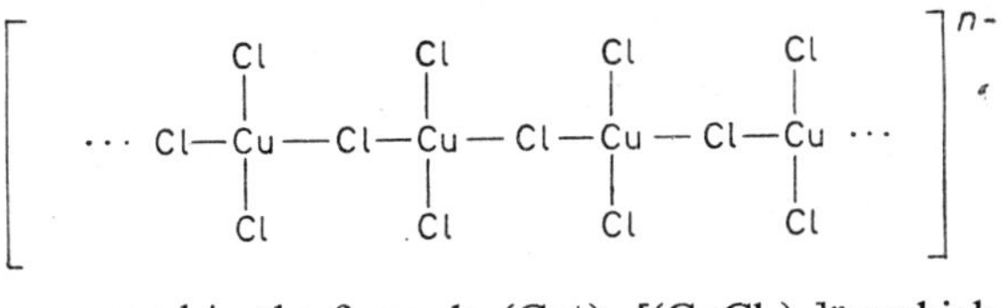

This may be expressed in the formula $(Cs^+)_n\ [(CuCl_3)_n]^{n-}$ which leads to the simple name caesium *catena*-μ-chloro-dichlorocuprate(II). If the structure were in doubt, however, the substance would be called caesium copper(II) chloride (as a double salt).

Examples:

1.

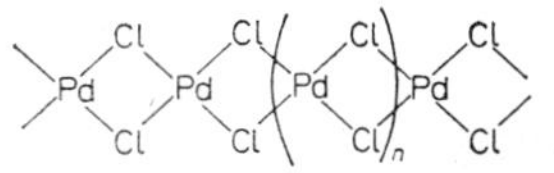

catena-di-μ-chloro-palladium

catena-di-μ-chloro-palladium(II)

2.

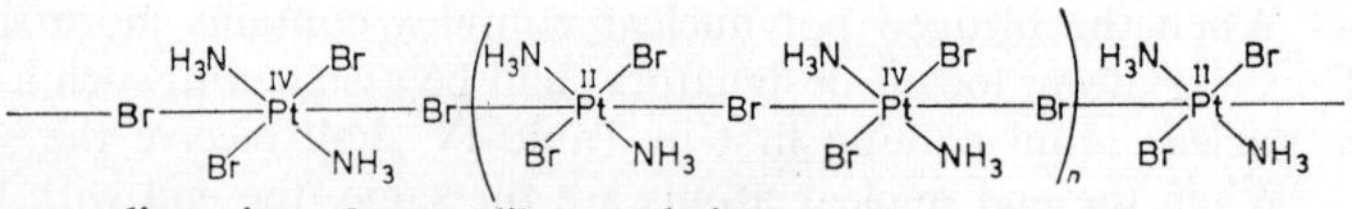

catena-diammine-μ-bromo-dibromoplatinum
catena-diammine-μ-bromo-dibromoplatinum(II, IV)

Chains may also be formed by the attachment of two centres of coordination to different atoms of the bridging group. In such instances appending the italicized symbols of the coordinating atoms to the name of the bridging ligand indicates the structure.

3.

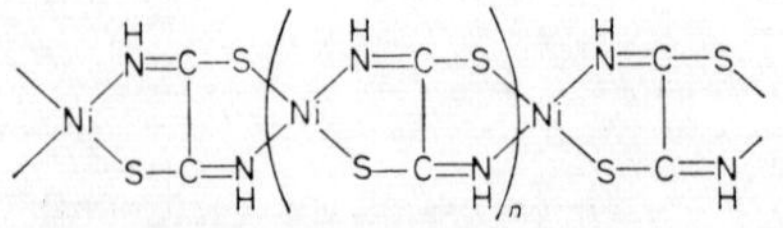

catena-μ-[dithiooxamidato(2−)-*N,S′*:*N′,S*]-nickel
catena-μ-[dithiooxamidato(2−)-*N,S′*:*N′,S*]-nickel(II)

4.

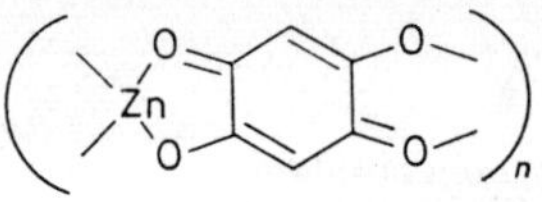

catena-μ-[2,5-dioxido-*p*-benzoquinone(2−)-*O,O′*:*O″,O‴*]-zinc
catena-μ-[2,5-dioxido-*p*-benzoquinone(2−)-*O,O′*:*O″,O‴*]-zinc(II)

7.7. DI- AND POLYNUCLEAR COMPOUNDS WITHOUT BRIDGING GROUPS

7.71—Direct Linking between Centres of Coordination

7.711—There are a number of compounds containing metal–metal bonds. Such compounds, when symmetrical, are named by the use of multiplicative prefixes; when unsymmetrical, one central atom and its attached ligands shall be treated as a ligand on the other central atom. The metal to be considered as the primary central atom is the last encountered in Table IV.

The names for organometallic radicals, *e.g.*, chloromercurio and dimethylarsenio are constructed by prefixing the names of the organic and inorganic radicals to the modified name of the metal given in Table V.

Examples:

1. $[Br_4Re—ReBr_4]^{2-}$
bis(tetrabromorhenate)(2−)
bis[tetrabromorhenate(III)]
2. $[(CO)_5Mn—Mn(CO)_5]$
bis(pentacarbonylmanganese)
3. $[(CO)_4Co—Re(CO)_5]$
pentacarbonyl(tetracarbonylcobaltio)rhenium
4. $[\eta\text{-}C_5H_5(CO)_3Mo—Mo(CO)_3\text{-}\eta\text{-}C_5H_5]$
bis(tricarbonyl-η-cyclopentadienylmolybdenum)
5. $[(Cl_3Sn)_2RhCl_2Rh(SnCl_3)_2]^{4-}$
di-μ-chloro-bis[bis(trichlorostannyl)rhodate](4−) ion
di-μ-chloro-bis[bis(trichlorostannyl)rhodate(I)] ion
6. $[(C_6H_5)_3AsAuMn(CO)_5]$
pentacarbonyl[(triphenylarsine)aurio]manganese

7. $[\{o\text{-}(CH_3)_2AsC_6H_4\}_2(CH_3)AsAgCo(CO)_4]$
   {bis[2-(dimethylarsino)phenyl]methylarsine}argentiotetracarbonylcobalt
8. $[\eta\text{-}C_5H_5(CO)_3W\text{—}Mo(CO)_3\text{-}\eta\text{-}C_5H_5]$
   tricarbonyl-η-cyclopentadienyl(tricarbonyl-η-cyclopentadienylmolybdio)tungsten
9. $[\{(C_6H_5)_3P\}_2CO(Cl)_2Ir\text{—}HgCl]$
   carbonyldichloro(chloromercurio)bis(triphenylphosphine)iridium

7.712—Where there are bridging groups as well as the metal–metal bond between the same pair of atoms, the compound is named as a bridged compound. When necessary or desired the existence of the the metal–metal bond is indicated by the italicized symbols in parentheses at the end of the name.

Examples:

1. $(OC)_3Co(CO)_2Co(CO)_3$
   di-μ-carbonyl-bis(tricarbonylcobalt)(*Co—Co*)
2. $(\eta\text{-}C_5H_5)NiPhC{\equiv}CPhNi(\eta\text{-}C_5H_5)$
   μ-(diphenylacetylene)-bis(η-cyclopentadienylnickel)(*Ni—Ni*)
3. $\eta\text{-}C_8H_{10}(CO)Co(CO)_2Co(CO)\text{-}\eta\text{-}C_8H_{10}$
   di-μ-carbonyl-bis[carbonyl-η-(1,3,6-cyclooctatriene)cobalt](*Co—Co*)
4. $(CO)_3Fe(C_2H_5S)_2Fe(CO)_3$
   bis(μ-ethylthio)-bis(tricarbonyliron)(*Fe—Fe*)
5. $[\{Fe(CO)_3\}_3(CO)_2]^{2-}$
   di-μ_3-carbonyl-*cyclo*-tris(tricarbonylferrate)(3 *Fe—Fe*)(2−)
6. $[(Ni\text{-}\eta\text{-}C_5H_5)_3(CO)_2]$
   di-μ_3-carbonyl-*cyclo*-tris(cyclopentadienylnickel)(3 *Ni—Ni*)
7. $Os_3(CO)_{12}$
   cyclo-tris(tetracarbonylosmium)(3 *Os—Os*)
   (For alternate name see **7.72**, example 1.)
8. 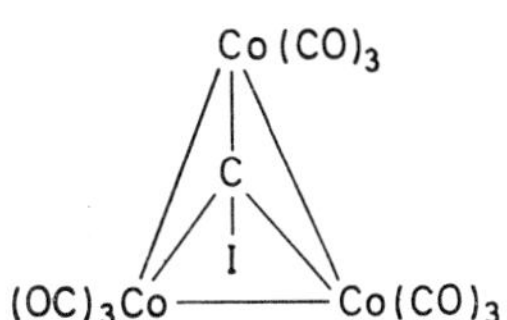

   μ_3-iodomethylidyne-*cyclo*-tris(tricarbonylcobalt)(3 *Co—Co*)
9. Ph Ph C≡C OC CO OC—Co—Co—CO OC CO

   hexacarbonyl-μ-η-(diphenylacetylene)-dicobalt(*Co—Co*)

7.72—Homoatomic Aggregates

There are several instances of a finite group of metal atoms with bonds directly between the metal atoms but also with some nonmetal atoms or groups (ligands) intimately associated with the *cluster*. The geometrical shape of the cluster is designated by *triangulo*, *quadro*, *tetrahedro*, *octahedro*, *etc.*, and the nature of the bonds to the ligands by the conventions for bridging bonds and simple bonds. Numbers are used as locant designators as they are for homoatomic chains and boron clusters (*cf.* **11**).

Examples:

1. $Os_3(CO)_{12}$
dodecacarbonyl-*triangulo*-triosmium
(For alternate name see **7.712**, example 7)
2. $Cs_3[Re_3Cl_{12}]$
caesium dodecachloro-*triangulo*-trirhenate(3−)
tricaesium dodecachloro-*triangulo*-trirhenate
3. B_4Cl_4
tetrachloro-*tetrahedro*-tetraboron
4. $[Nb_6Cl_{12}]^{2+}$
dodeca-μ-chloro-*octahedro*-hexaniobium(2+) ion
5. $[Mo_6Cl_8]^{4+}$
octa-μ_3-chloro-*octahedro*-hexamolybdenum(4+) ion
octa-μ_3-chloro-*octahedro*-hexamolybdenum(II) ion
6. $[Mo_6Cl_8Cl_6]^{2-}$
octa-μ_3-chloro-hexachloro-*octahedro*-hexamolybdate(2−) ion
octa-μ_3-chloro-hexachloro-*octahedro*-hexamolybdate(II) ion
7. $[Mo_6Cl_8Cl_3\{(C_6H_5)_2PCH_2CH_2P(C_6H_5)_2\}py]$ Cl
octa-μ_3-chloro-trichloro[ethylenebis(diphenylphosphine)]-pyridine-*octahedro*-hexamolybdenum(1+) chloride
octa-μ_3-chloro-trichloro[ethylenebis(diphenylphosphine)]-pyridine-*octahedro*-hexamolybdenum(II) chloride
8. B_8Cl_8
octachloro-*dodecahedro*-octaboron

This system is readily adaptable to naming compounds which are difficult to name by other patterns.

9.

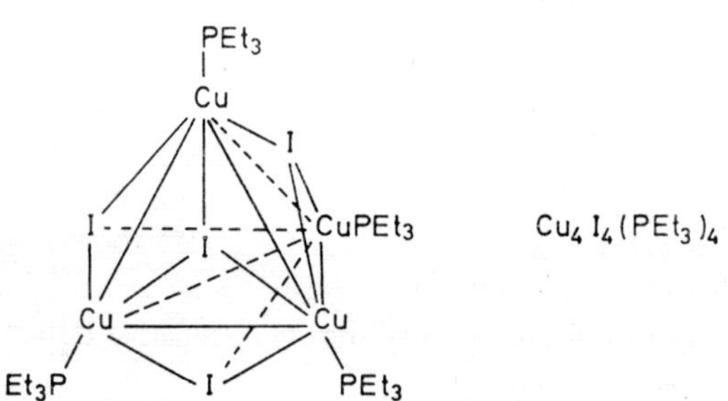

tetra-μ_3-iodo-tetrakis(triethylphosphine)-*tetrahedro*-tetracopper
tetra-μ_3-iodo-tetrakis(triethylphosphine)-*tetrahedro*-tetracopper(I)

10.

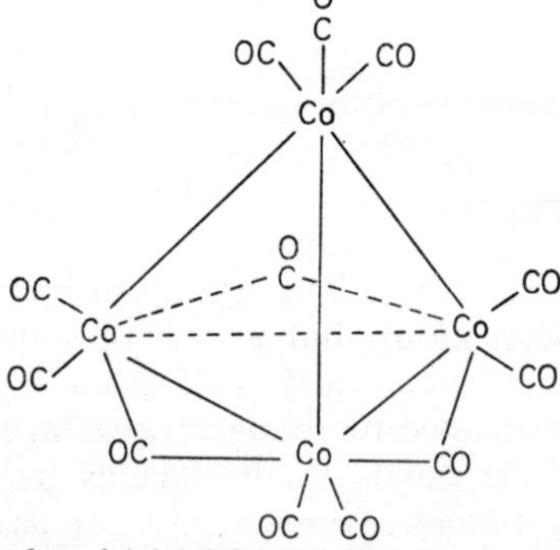

2,3;3,4;4,2-tri-μ-carbonyl-1,1,1-2,2,3,3,4,4-nonacarbonyl-*tetrahedro*-tetracobalt

11.

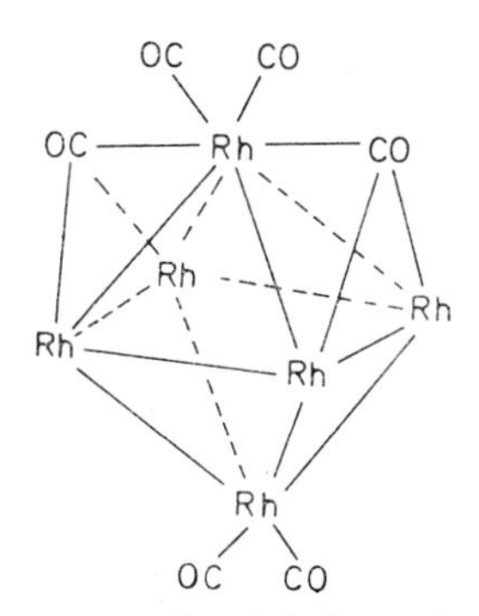

1,2,3;1,4,5;2,5,6;3,4,6-tetra-μ_3-carbonyl-dodecacarbonyl-*octahedro*-hexarhodium

The non-bridging CO groups of the lowest rhodium atom and the complete surroundings of the rhodium atom at the top are shown. Other CO groups are omitted to avoid confusion in the diagram. The whole structure is such that all rhodium atoms are equivalent. There are two additional bridging CO groups, one above the octahedron face formed by atoms 2, 5 and 6 and another above that formed by atoms 3, 4 and 6. Every rhodium atom has two non-bridging CO groups.

7.8. ABSOLUTE CONFIGURATIONS CONCERNED WITH SIX-COORDINATED COMPLEXES BASED ON THE OCTAHEDRON*

Introduction

Configuration. For spectroscopic purposes and for following the stereochemical course of substitution reactions it is of interest to consider, for example, tris- and bis-bidentate six-coordinated complexes based on the octahedron as related through the configurations depicted in *Figure 1* (a) and (b). Here the edges spanned by the chelate rings are drawn as heavy lines. The chelate rings are thought of as devoid of chemical significance in the sense that the chelating ligands may be identical or different, and may be symmetrical or not. Similarly the two X's represent two unidentate ligands which may or may not be identical. It is desired, in all generality, to have a designation of chirality which is independent of the chemical nature of the chelating ligands

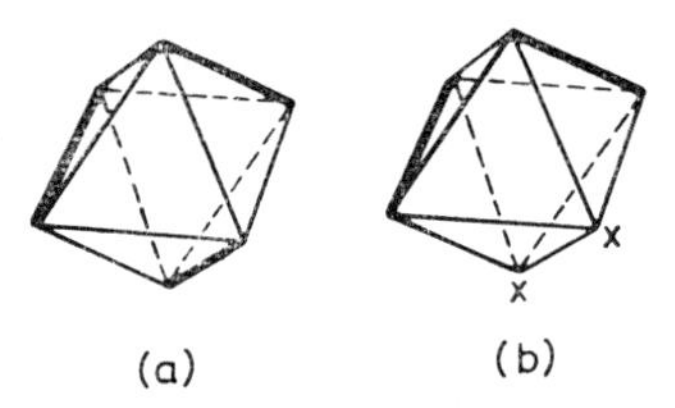

Figure 1. General "octahedral" systems containing three (a) and two (b) bidentate ligands represented by the edges (drawn as heavy lines) which they span. It is desired to characterize these systems as having the same absolute configuration independently of their chemical significance. They have both been designated by Δ in the present proposal.

* The rules here are given in short form in **7.87**.

and which only depends on the relative positions of the heavy line edges which represent the bidentate ligands or the bidentate units of multidentate ligands. *Conformation.* Further, for spectroscopic purposes it is of interest to designate the conformation of chelate rings relative to the central atom or ion, but independently of the other atoms forming the chelate ring and also of the substituents of these atoms.

The present proposals. All the rules which follow are based on the fact that two skew and non-orthogonal lines define a helical system. They primarily describe a nomenclature for the absolute configuration of classes comprising *cis*-bis-bidentate and tris-bidentate complexes and the absolute conformation of five-membered chelate rings. However, since the rules are based on general grounds, they lend themselves readily to application to more complicated situations, *i.e.*, multidentate chelate systems and larger chelate rings.

In the chemical literature there exist different proposals for the nomenclature of the systems which are under consideration here. These proposals are devoid of chemical significance and generally based upon helicities about symmetry or pseudo-symmetry axes. The present proposals are independent of symmetry concepts and thereby easier to generalize to situations where symmetry is absent.

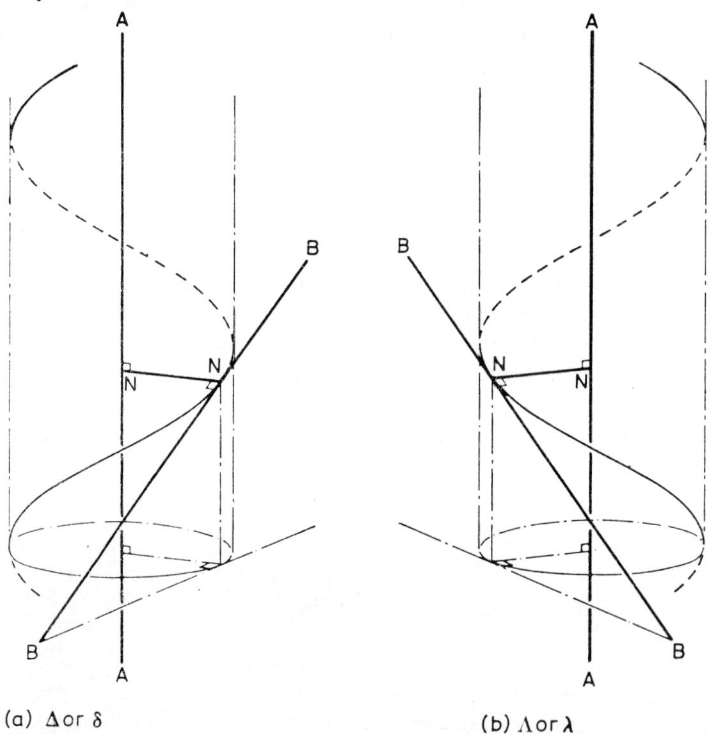

Figure 2. Two skew lines *AA* and *BB* which are not orthogonal define a helical system. In the figure *AA* is taken as the axis of a cylinder whose radius is determined by the common normal *NN* of the two skew lines. The line *BB* is a tangent to the above cylinder at its crossing point with *NN* and defines a helix upon this cylinder by being the tangent to it at this crossing point. (a) and (b) illustrate a right- and a left-handed helix.

7.81—Basic Principle

Two skew lines which are not orthogonal to each other make up a helical system as illustrated in *Figures 2* and *3*. Two skew lines possess the property of having one and only one normal in common. In *Figure 2* one of the skew lines *AA* determines the axis of a helix upon a cylinder whose radius is equal to the length of the two skew lines' common normal *NN*. The other of the skew lines *BB* makes up a tangent to the helix at *N* and determines the steepness of the helix. In *Figure 3* the two skew lines *AA* and *BB* are seen in projection on to a plane orthogonal to their common normal.

(a) of *Figure 2* and *3* illustrates a right-handed helix to be associated with the Greek letter delta (Δ referring to configuration, δ to conformation). (b) of *Figures 2* and *3* illustrates a left-handed helix to be associated with the Greek letter lambda (Λ for configuration, λ for conformation)*.

Because we are only interested in a qualitative measure of the helicity, the steepness of a helix is, in general, of no importance. However, the singularities at infinite steepness, where the skew lines become parallel lines, and at vanishing steepness, where the lines become orthogonal, should be noted. Here an infinitely small rotation of one line relative to the other about their common normal will change the helicity from right-handedness to left-handedness or *vice versa*. It is obvious that as the representation of our physical situation approaches these singularities the helicity becomes undefined (see *Figure 13*).

7.82—Application to Configuration

7.821—*Representation of chelate rings*. A chelate ring of a six-coordinated complex, whose ligators form an approximate octahedron, is represented by the edge determined by its two ligators. If two such edges are skew the pair can, without any further conventions, be associated† with either (a) or (b) of *Figure 3*. This is the basis of the present proposal for nomenclature of absolute configurations, for *cis*- bis-bidentate and tris-bidentate systems.

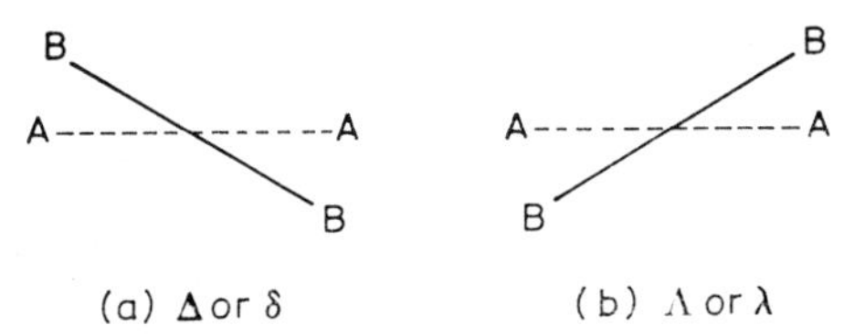

Figure 3. The figure shows pairs of non-orthogonal skew lines in projection upon a plane parallel to both lines. The fully drawn line *BB* is above the plane of the paper, the dotted line *AA* below this plane. (a) corresponds to (a) of *Figure 2* and defines a right-handed helix. b) corresponds to (b) of *Figure 2* and defines a left-handed helix.

* It should be noted that orthogonal to the common normal of the two skew lines there is a two-fold axis (in fact, there are two such axes) of proper rotation which brings each one of the skew lines into coincidence with the other. This means that the helix which the first line, *BB*, say, determines around the second one, *AA*, has the same helicity as that which the second one determines around the first one.

† In connection with the singularities mentioned above it should be noted that by moving the ligators away from the ideal octahedral positions a gradual transition from the situation of *Figure 3* (a) to that of *Figure 3* (b) is possible without a change of absolute configuration. However, for this to occur the distortions must be so great that one would no longer think of calling the complex octahedral. Furthermore such cases are unknown.

Two heavy line edges which are neither neighbouring edges having a common vertex, nor opposite edges, will in an octahedron form a pair of skew lines. This pair always has the same relative position as that of a *cis*-bis-bidentate complex.

In *Figure 4* is seen the representation of the *cis*-bis-bidentate complex of *Figure 1* (b) redrawn so as to conform to *Figure 3*, and for the particular absolute configuration to *Figure 3* (a). To the corresponding tris complex [*Figure 1* (a)] is attributed the same designation because its three heavy line edges are equivalent and therefore also the three possible pairs of heavy line edges. This is illustrated in *Figure 5*.

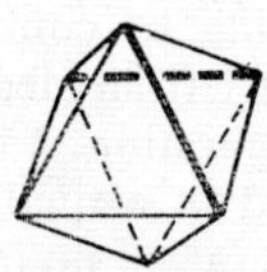

Figure 4. The bis-bidentate complex of *Figure 1* (b) redrawn so as to become associated with *Figure 3* (a) and thus to become designated by Δ.

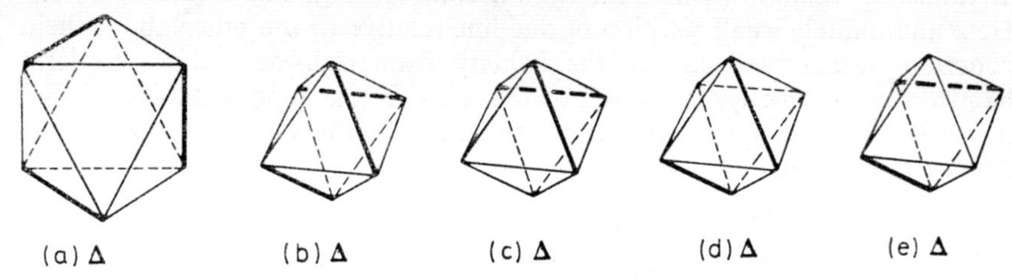

Figure 5. (a) and (b) show the tris-bidentate system of *Figure 1* (a) redrawn in two different ways. Since each of the bidentate ligands has lost its individuality by being represented only by the edge which it spans, the threefold axis of symmetry of the octahedron applies also to the present system. (a) shows the system in projection on a plane orthogonal to its threefold axis. (c), (d), and (e) each illustrates one of the three possible pairs of bidentate ligands oriented so as to refer to (b). (c) is associated with *Figure 3* (a), and thus is designated by Δ. The same must hold true also for (d) and (e) because the threefold axis makes the three pairs of representations of bidentate ligands equivalent.

7.822—*Multidentate systems*. It is straightforward to extend the application of the above rules to more complex situations involving multidentate ligands. It is by analogy with the tris-bidentate case (*Figure 5*) only a matter of studying the interrelations between all the chelate rings whose corresponding edges form a pair of skew lines, *i.e.*, all the ring pairs whose relative position is the same as in a *cis*-bis-bidentate complex. Now one might count up all such contributions and designate the complex situation by Δ if the number of Δ contributions from the individual pairs exceeds the number of Λ contributions and *vice versa*. This convention, which could be applied to the situations shown in *Figures 6–8*, will *not* be recommended here for the reason given in the next paragraph. Even though non-helical situations will always contribute Δ and Λ an equal number of times (*Figure 9*), the same may be true as well or certain helical situations, as illustrated in *Figure 10*.

A case such as that of *Figure 10* requires a further convention and here no simple one has yet been proposed. A possible convention here might conflict with the above simple counting of Δ and Λ contributions. We therefore recommend, at the present stage, for the case of *Figures 6–8* where the number of Δ and Λ contributions is different, to characterize the complexes as follows: *Figure 6*, "skew chelate pair, Δ"; *Figure 7*, "skew chelate pair, Λ"; *Figure 8*, "skew chelate pairs, $\Lambda\Delta\Lambda$". In the last example the order of the Greek letter symbols is immaterial. The case of *Figure 10* might at present be characterized by "the end chelate rings form a skew chelate pair, Λ".

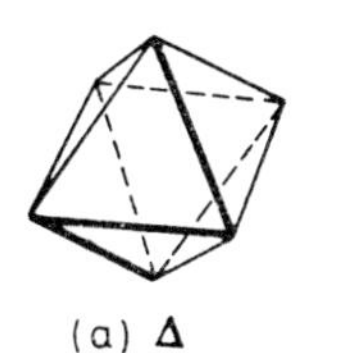

Figure 6. A quadridentate system (a). Here only two of the heavy line edges are skew, the pair (b) being associated with *Figure 5* (d) and thus being designated by Δ. The system as a whole is proposed designated "skew chelate pair, Δ". The system may be thought of as representing the α-isomer of a trien-complex (trien = triethylenetetraamine).

Figure 7. Another quadridentate complex (a). As in *Figure 6* there is only one helical pair (b). This is clearly associated with the mirror image of *Figure 5* (c) (and, of course, therefore, also of *Figure 5* (d) and (e), although this is less easy to see) and thereby gives rise to the designation Λ. The system is therefore designated "skew chelate pair, Λ". The system may represent the β-isomer of a trien-complex (see *Figure 6*).

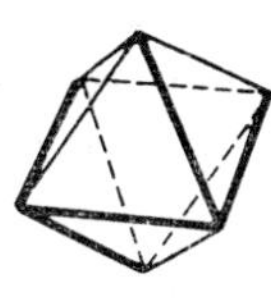

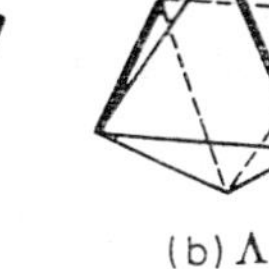

Figure 8. A sexidentate complex (a). The pair (c) is associated with *Figure 5* (d) and is therefore Δ. The pairs (b) and (d) are clearly the mirror images of *Figure 5* (c) and *5* (e), respectively, and are therefore both Λ. The whole system is designated "skew chelate pairs, $\Lambda\Delta\Lambda$" where the order of the symbols is immaterial. The system may represent an edta complex (edta = ethylenediaminetetraacetate).

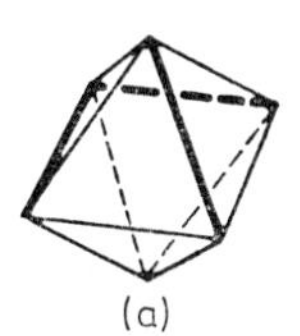

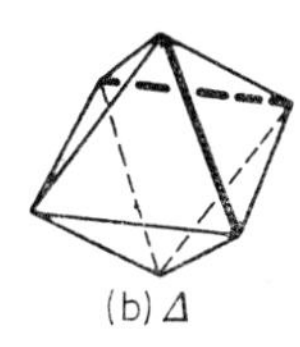

Figure 9. A non-helical system (a). The helical pairs (b) and (c) are mirror images of each other and contribute Δ and Λ, respectively. Non-helical systems always have an equal number of Δ and Λ contributions. The reverse conclusion, however, is not valid (see *Figure 10*).

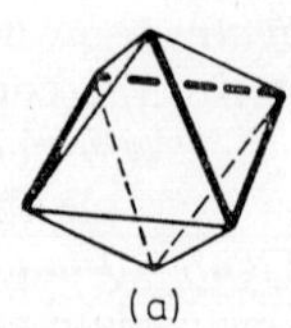

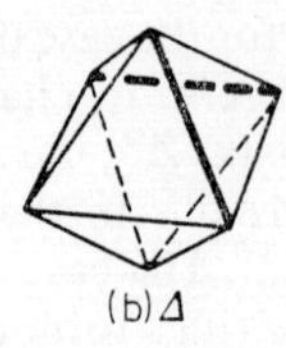

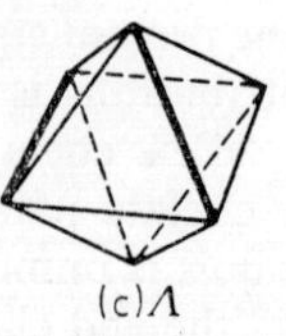

Figure 10. A quinquedentate system (a). (b) is Δ by association with *Figure* 5(c), (c) is Λ because it is the mirror image of (b). A designation for the whole helical system (a) cannot be obtained without a further convention. A preliminary designation might be "the end chelate rings form a skew chelate pair, Λ".

7.83—Application to Conformation

In order to define the helicity of a ring conformation a convention is required for making a choice of a pair of skew lines. Here it is proposed to choose one of the lines of this pair as the edge covered by the chelate ring, *i.e.*, the line *AA* joining the two ligators. The other line *BB* is taken as that joining the two ring atoms which are neighbours to each of the ligators.

Two enantiomeric situations are shown in projection in *Figure 11*. The two ligators *AA* are in the plane of the paper, the central atom *M* is below this plane and the two neighbouring ring atoms *BB* are above it. *Figure 11* (a) and (b) are associated with the corresponding *Figure 3*, and the proposed convention for designating the helicity is thereby given. In *Figure 12* is shown a situation to which is attributed the same designation as that of the case of *Figure 11* (a). In *Figure 13 BB* is parallel to *AA* and the chelate ring will not be helical at least up to a ring size of seven or eight members, which for our purpose is without importance. The situation in which *BB* is parallel to *AA* corresponds to the case of any planar chelate ring, and in addition to this, for a five-membered ring, it corresponds to the envelope form, and for the six-membered ring, either to the chair or to the boat form. In this case only the skew-boat form has a helical character.

Non-helical situations may still represent chiral situations when the chemical significance of the atoms, *i.e.* their possibility of being different, is considered. The present nomenclature problem, however, is not concerned with such cases.

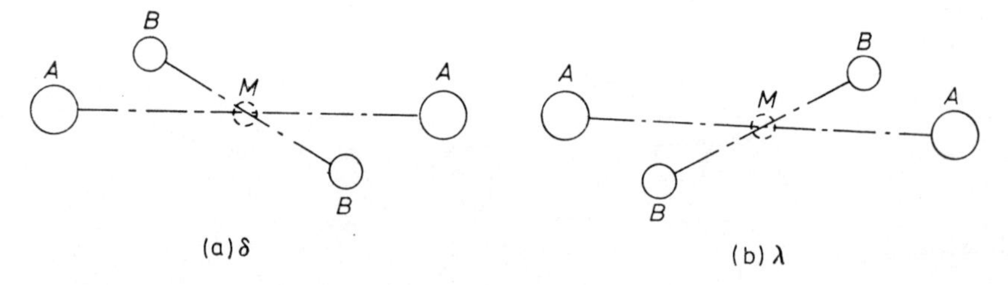

Figure 11. Illustration of the convention for designating the helical character of the conformation of chelate rings. The ligating atoms in the plane of the paper determine one of the skew lines *AA*. The neighbouring atoms of each ligator determine the other line *BB*, which here is above the plane of the paper, the central atom *M* being below this plane. The designations become clear by comparison with *Figure 3* (a) and (b).

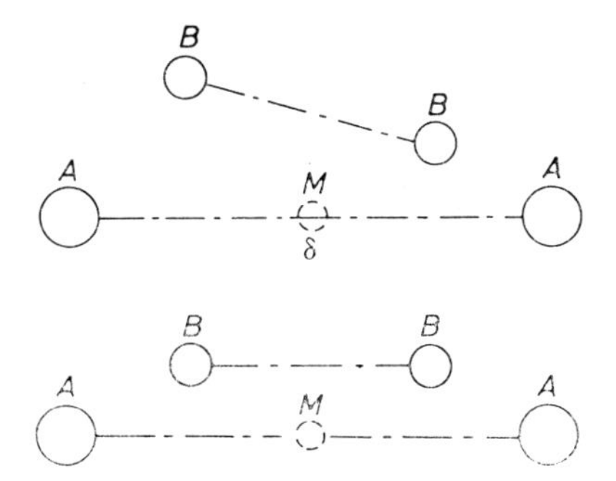

Figure 12. Illustration of an alternative situation to that of *Figure 11* (a). Both atoms *BB* are above the plane determined by *M* and *AA*, but this is immaterial from a nomenclature point of view. The lines *AA* and *BB* are still skew and correspond to the situation of *Figure 3* (a).

Figure 13. Non-helical chelate ring drawn as in *Figure 11*. This figure illustrates a five-membered ring in its envelope form or a six-membered ring in either its boat or its chair form.

7.84—Absolute Configurations

The proposals which have been put forward here dictate that the absolute configuration of the tris(ethylenediamine)cobalt(III) ion with a positive rotation at the Na_D line be characterized as upper case Λ and (—)propylenediamine in its stable chelate conformer (equatorial CH_3-) be characterized by lower case λ.

7.85—Phenomenological Characterization

As well as the symbols for designating structure some phenomenological description of a mirror-image isomer is essential. The isomer might be denoted by its sign of rotation at a particular wavelength, $(+)_\lambda$,

e.g., $(+)_{589}[Co(en)_3]\ Cl_3$ en = ethylenediamine (see **7.35**)

When optically active ligands are coordinated, they are denoted as (+) and (—) where the signs are the signs of rotation of the ligand at the Na_D line,

e.g., $(-)_{589}[Co\{(-)pn\}_3]\ Cl_3$

In those instances where the absolute configuration of the ligand is known this might also be included in the description,

e.g., $(-)_{589}[Co\{(R)(-)pn\}_3]\ Cl_3$

7.86—Full Characterization

Examples of the use of the full nomenclature proposed here follow:

$\Lambda\ (+)_{589}[Co\{(+)pn\}_2\{(-)pn\}\delta\delta\lambda]\ Cl_3$

$\Delta\ (-)_{589}[Co\{(R)\ (-)pn\}_3\lambda\lambda\lambda]\ (+)_{546}[Rh(C_2O_4)_3]$

7.87—Designation of Configurational Chirality Caused by Chelation in Six-coordinated Complexes Based on the Octahedron

7.871—*Cis-bis-bidentate chelation.* The two ligating atoms of a chelate ring define a line. Two such lines for the pair of chelate rings define a helix. One line is the axis of the helix and the other is the tangent of the helix at the common normal for the skew lines. The tangent describes a right-handed (Δ) or a left-handed (Λ) helix with respect to the axis and thereby defines the configuration.

7.872—*Tris-bidentate chelation.* Any one of the three pairs of chelate rings is chosen to designate the configuration by **7.871**.

7.873—*Multidentate chelation.* Chiral complexes of multidentate ligands are considered to contain pairs of skew lines (**7.871**) and are designated by all the symbols, Δ's and Λ's, belonging to all the skew-line pairs. The order of citation of the symbols is immaterial.

7.88—Designation of Conformational Chirality of a Chelate Ring

The line joining the two ligating atoms and the line joining the two atoms of the chelate ring adjacent to each of the ligating atoms define a helix. One line is the axis of the helix and the other is the tangent of the helix at the common normal for the skew lines. The tangent describes a right-handed (δ) or a left-handed (λ) helix with respect to the axis and thereby defines the conformation.

APPENDIX

Relationships Between the Proposed Symbols and Those in Earlier Use

The symbols Δ and Λ were originally proposed for tris-bidentate complexes by PIPER[1] who used the threefold axis (C_3) as reference axis. The present convention agrees with the results of PIPER's proposal. The present convention for designation of conformation likewise agrees with LIEHR's[2] proposed use of δ and λ.

The absolute configuration[3] of $(+)_{589}[Co(en)_3]^{3+}$ (in the crystal, $\Lambda\ \delta\delta\delta$) is Λ and that[4] of $(-)_{589}[Co\{(-)pn\}_3]^{3+}$ is $\Delta\ \lambda\lambda\lambda$, as determined by SAITO *et al.* from X-ray crystallography. These two complex ions were in the X-ray papers designated by *D* and *L*, respectively. MASON pointed out that the helical configuration about a C_2 axis of a tris-bidentate or *cis*-dianionobis-bidentate complex is opposite to that about the respective C_3 or pseudo C_3 axes. He proposed[5] the use of *P* (positive for right-handed) and *M* (minus for left-handed) as in $P(C_3)$ or $M(C_2)$ where the reference axis is indicated. The result of the present proposal is equivalent to that using the C_3 or pseudo C_3 axis and is *opposite* to that using a C_2 axis, *i.e.*, Δ for $P(C_3)$ or $M(C_2)$ and Λ for $M(C_3)$ or $P(C_2)$. HAWKINS and LARSEN[6] defined an octant sign to characterize the helicity of configurations (also of multidentate systems) as well as for conformations. For tris-bidentate and *cis*-bis-bidentate systems and for conformations of five and six-membered rings the relation to the present proposal is Λ (positive octant sign), λ (negative octant sign). LEGG and DOUGLAS[7] suggested the general use of the C_2 axis for reference of helicity and a ring-pairing method for assigning the helicity of complexes containing multidentate ligands. The ring pairs chosen to define the helicity are the same as those proposed here. However, because of the C_2-reference axis their use of Δ and Λ is opposite to that of the present proposal. It should further be noted that both the octant-sign method and the ring-pairing method of characterizing absolute configurations need extra conventions in certain cases of the type discussed here along with *Figure 10*.

COREY and BAILAR[8] and SARGESON[9] have discussed the concomitant interplay of conformation and configuration in tris-bidentate diamine complexes. These authors designated the conformation of the five-membered ethylenediamine ring as k and k', but used k and k' in the opposite sense*. With reference to our *Figure 11* the interrelation of δ, λ and k, k' is

	COREY and BAILAR	SARGESON
δ	k'	k
λ	k	k'

Acknowledgment. The Commission wishes to express its appreciation of the valuable help offered by Dr. WERNER FENCHEL, Professor of Mathematics in the University of Copenhagen, by the late Sir CHRISTOPHER INGOLD and Dr. R. S. CAHN, and by several chemists working with optically active complexes; they are especially indebted to Drs. B. E. DOUGLAS, A. SARGESON and C. E. SCHÄFFER who attended meetings of the Commission.

* The cause of the confusion with respect to k and k' is an error in the upper drawing of *Figure 3* of COREY and BAILAR's paper! The ring conformations of the unstable form, the *ob* form $\Delta\ \delta\delta\delta$, is correctly given as $k'k'k'$ in the lower drawing of their *Figure 3*, but the stable form, the *lel* form $\Delta\ \lambda\lambda\lambda$, discussed in the text correctly as kkk, appears in the upper drawing of their *Figure 3* as $\Delta\ \delta\delta\delta$.

References

1 T. S. PIPER. *J. Amer. Chem. Soc.* **83**, 3908 (1961)
2 A. D. LIEHR. *J. Phys. Chem.* **68**, 3629 (1964)
3 Y. SAITO, K. NAKATSU, M. SHIRO and H. KUROYA. *Acta Cryst.* **8**, 729 (1955); K. NAKATSU, M. SHIRO, Y. SAITO and H. KUROYA. *Bull. Chem. Soc. Japan* **30**, 158 (1957)
4 Y. SAITO, H. IWASAKI and H. OTA. *Bull. Chem. Soc. Japan* **36**, 1543 (1963)
5 A. J. MCCAFFERY, S. F. MASON and R. E. BALLARD. *J. Chem. Soc.* 2883 (1965); A. J. MCCAFFERY, S. F. MASON and B. J. NORMAN. *J. Chem. Soc.* 5094 (1965)
6 C. J. HAWKINS and E. LARSEN. *Acta Chem. Scand.* **19**, 185 and 1969 (1965)
7 J. I. LEGG and B. E. DOUGLAS. *J. Amer. Chem. Soc.* **88**, 2697 (1966)
8 E. J. COREY and J. C. BAILAR, JR. *J. Amer. Chem. Soc.* **81**, 2620 (1959)
9 A. SARGESON. *Transition Metal Chemistry*, Ed. R. L. Carlin, Marcel Dekker, New York, **3**, 303 (1966)

8. ADDITION COMPOUNDS

This rule covers some donor-acceptor complexes and a variety of lattice compounds. It is particularly relevant to compounds of uncertain structure; new structural information often makes naming according to Section 7 possible.

The ending -ate is now the accepted ending for *anions* generally and should not be used for addition compounds. Alcoholates are the *salts* of alcohols and this name should not be used to indicate alcohol of crystallization. Analogously addition compounds containing ether, ammonia, *etc.*, should *not* be termed etherates, ammoniates, *etc.*

However, one exception has to be recognized. According to the commonly accepted meaning of the ending -ate, "hydrate" would be, and was formerly regarded as, the name for a *salt* of water, *i.e.* what is now known as a hydroxide; the name hydrate has now a very firm position as the name of a compound containing water of crystallization and is allowed also in these Rules to designate water bound in an unspecified way; it is preferable, however, even in this case to avoid the ending -ate by using the name "water" (or its equivalent in other languages) when possible.

The names of addition compounds may be formed by connecting the names of individual compounds by spaced hyphens and indicating the number of molecules after the name by Arabic numerals separated by the solidus. Boron compounds and water are always cited last in that order. Other molecules are cited in order of increasing number; any which occur in equal numbers are cited in alphabetical order.

Examples:

	Formula	Name
	Solvates and Molecular compounds	
1.	$3CdSO_4 \cdot 8H_2O$	cadmium sulfate – water (3/8)
2.	$Na_2CO_3 \cdot 10H_2O$	sodium carbonate – water (1/10) or sodium carbonate decahydrate
3.	$Al_2(SO_4)_3 \cdot K_2SO_4 \cdot 24H_2O$	aluminium sulfate – potassium sulfate – water (1/1/24)
4.	$CaCl_2 \cdot 8NH_3$	calcium chloride – ammonia (1/8)
5.	$AlCl_3 \cdot 4C_2H_5OH$	aluminium chloride – ethanol (1/4)
6.	$2CH_3OH \cdot BF_3$	methanol – boron trifluoride (2/1)
7.	$NH_3 \cdot BF_3$	ammonia – boron trifluoride (1/1)
8.	$BiCl_3 \cdot 3PCl_5$	bismuth trichloride – phosphorus pentachloride (1/3)
9.	$TeCl_4 \cdot 2PCl_5$	tellurium tetrachloride – phosphorus pentachloride (1/2)
10.	$BF_3 \cdot 2H_2O$	boron trifluoride – water (1/2)
	Clathrates	
11.	$8H_2S \cdot 46H_2O$	hydrogen sulfide – water (8/46)
12.	$8Kr \cdot 46H_2O$	krypton – water (8/46)
13.	$6Br_2 \cdot 46H_2O$	bromine – water (6/46)
14.	$8CHCl_3 \cdot 16H_2S \cdot 136H_2O$	chloroform – hydrogen sulfide – water (8/16/136)
15.	$C_6H_6 \cdot NH_3 \cdot Ni(CN)_2$	ammonia – benzene – nickel(II) cyanide (1/1/1)

Part of the adduct can often be named according to **7.2** and **7.3**, particularly if structural information is available.

Examples:

16. $[Fe(H_2O)_6] SO_4 \cdot H_2O$ hexaaquairon(II) sulfate monohydrate
17. $[(CH_3)_4N] [AsCl_4] \cdot 2AsCl_3$ tetramethylammonium tetrachloroarsenate(III)-arsenic trichloride (1/2)

9. CRYSTALLINE PHASES OF VARIABLE COMPOSITION

Compounds Involving Isomorphous Replacement, Interstitial Solutions, Intermetallic Compounds, Semiconductors and other Nonstoicheiometric Compounds (Berthollides)

9.11—If an intermediate crystalline phase (whether stable or metastable) occurs in a two-component (or more complex) system, it may obey the law of constant composition with very high accuracy, as in the case of sodium chloride, or it may be capable of varying in composition over an appreciable range, as occurs for example with FeS. A substance showing such a variation is called a *berthollide*.

In connection with the berthollides the concept of a characteristic or ideal composition is frequently used. A unique definition of this concept seems to be lacking, but usually the definition is based upon the crystal structure. Sometimes one can state several characteristic compositions. In spite of this the concept of a characteristic composition can be used when establishing a system of notation for phases of variable composition. It is also possible to use the concept even if the characteristic composition is not included in the known homogeneity range of the phase.

9.12—For the present, formulae should preferably be used for berthollides and solid solutions, since strictly logical names tend to become inconveniently cumbersome. The latter should only be used when unavoidable (*e.g.*, for indexing), and may be made in the style of: iron(II) sulfide (iron deficient); molybdenum dicarbide (excess of carbon), or the like. Mineralogical names should only be used to designate actual minerals and not to define chemical composition; thus the name calcite refers to a particular mineral (contrasted with other minerals of similar composition) and is not a term for the chemical compound whose composition is properly expressed by the name calcium carbonate. (The mineral name may, however, be used to indicate the structure type, see **6.52**).

9.21—Various notations are used for the berthollides, depending upon how much information is to be conveyed.

A general notation, which can be used even when the mechanism of the variation in composition is unknown, is to put the sign $\approx$ (read as *circa*) before the formula. (In special cases it may also be printed above the formula.)

Examples:

$\approx$FeS $\qquad \overset{\approx}{\text{CuZn}}$

If it is desirable to give more information, one of the following notations may be used.

9.22—For a phase where the variable composition is solely or partially caused by replacement, atoms or atomic groups which replace each other are separated by a comma and placed together between parentheses.

If possible the formula ought to be written so that the limits of the homogeneity range are represented when one or other of the two atoms or groups is lacking. For example, the symbol (Cu,Ni) denotes the complete range from pure Cu to pure Ni; likewise K(Br,Cl) comprises the range from pure KBr to pure KCl. If only part of the homogeneity range is referred to, the major constituent should be placed first.

Substitution accompanied by the appearance of vacant positions (combination of substitutional and interstitial solution) may receive an analogous notation. For example $(Li_2,Mg)Cl_2$ denotes the homogeneous phase from LiCl to $MgCl_2$. The formula $Al_6(Al_2,Mg_3)O_{12}$ represents the homogeneous phase from the spinel Al_2MgO_4 ($= Al_6Mg_3O_{12}$) to the spinel form of Al_2O_3 ($= Al_6Al_2O_{12}$).

9.23—A more complete notation, which should always be used in more complex cases, may be constructed by indicating in a formula the variables which define the composition. Thus, a phase involving substitution atom for atom of A for B may be written $A_{m+x}B_{n-x}C_p$.

Examples:

Cu_xNi_{1-x} and KBr_xCl_{1-x}

In the case of the γ-phase of the Ag–Cd system, which has the characteristic formula Ag_5Cd_8, the Ag and Cd atoms can replace one another to some extent and the notation would be $Ag_{5\pm x}Cd_{8\mp x}$. For the plagioclases the notation will be $Ca_xNa_{1-x}Al_{1+x}Si_{3-x}O_8$ or $Ca_{1-y}Na_yAl_{2-y}Si_{2+y}O_8$.* This shows immediately that the total number of atoms in the unit cell is constant.

The commas and parentheses called for in **9.22** are not required in this case.

Interstitial or subtractive solution, whether combined with substitutional solution or not, can be shown in an analogous way. For example, the homogeneous phase between LiCl and $MgCl_2$ becomes $Li_{2x}Mg_{1-x}Cl_2$, showing that the anion lattice remains the same but that one vacant cation position appears for every replacement of $2Li^+$ by Mg^{2+}. The phase between Al_2MgO_4 and Al_2O_3 can be written $Al_6Al_{2(1-x)}Mg_{3x}O_{12}$ which shows that it cannot contain more Mg than that corresponding to Al_2MgO_4 ($x = 1$).

Further examples:

$Fe_{1-x}Sb$; $Fe_{1-x}O$; $Fe_{1-x}S$; $Cu_{2-x}O$; $Ca_xY_{1-x}F_{3-x}$

$Na_{1-x}WO_3$ or Na_yWO_3 (sodium tungsten bronzes, depending on the choice of characteristic composition).

For $x = 0$ each of these formulae corresponds to a characteristic composition. If it is desired to show that the variable denoted by x can only attain small values, this may be done by substituting δ or ϵ for x.

When using this notation, a particular composition can be indicated by stating the actual value of the variable x. Probably the best way of doing this is to put the value in parentheses after the general formula. For example, $Fe_{3x}Li_{4-x}Ti_{2(1-x)}O_6$ ($x = 0.35$). If it is desired to introduce the value of x into

* The plagioclases are aluminosilicates and aluminium belongs to the anionic part of the crystal lattice.

the formula itself, the substitution is more clearly understood if one writes $Fe_{3\times 0\cdot 35}Li_{4-0\cdot 35}Ti_{2(1-0\cdot 35)}O_6$ instead of $Fe_{1.05}Li_{3.65}Ti_{1.30}O_6$.

The solid solution of hydrogen in palladium can be written as PdH_x ($x<0.1$) and the palladium hydride phase as PdH_x ($0.5<x<0.7$). A phase of the composition M which has dissolved a variable amount of water can be written $M(H_2O)_x$.

9.31—If in addition to the chemical composition the existence of vacant sites and interstitial sites is to be shown, this can be done by using additional symbols as indicated in **9.311–9.314**.

9.311—A site in the structure of the ideal composition is represented by the square, □, and an interstitial site by the triangle, △. When it must be shown that a site is cationic or anionic, a cationic site is shown by □cat and an anionic site by □an. Crystallographically different sites can be distinguished by additional symbols *e.g.*, □a, □b, or □tet, □oct, the last two denoting tetrahedral or octahedral sites. A more precise notation could be obtained by putting in brackets immediately after the site symbol the point group symbol showing the symmetry of the immediate environment of the site and its coordination number, *e.g.*, $\square[O_h;6]$.

9.312—An atom A in the site □ is expressed by the symbol (A|□). Spinel can thus be represented by $(Al|\square oct)_2\ (Mg|\square tet)O_4$, Al being situated in octahedral and Mg in tetrahedral sites formed by the oxygen atoms. The "inverse spinel", magnetite, is represented by $(Fe^{II}{}_{\frac{1}{2}}Fe^{III}{}_{\frac{1}{2}}|\square oct)_2(Fe^{III}|\square tet)$ O_4, which means that Fe^{II} and half of the Fe^{III} are distributed at random over certain octahedral sites.

9.313—If *n* atoms A are distributed over *m* sites □, this is expressed by $(A_n|\square_m)$. This implies that $m-n$ sites are vacant, and it is not necessary to show them specially. The γ-modification of Fe_2O_3 is thus $(Fe^{III}{}_{\frac{8}{3}}|\square_3)O_4$. Using this notation we can write lithium magnesium chloride as $(Li_{2x}, Mg_{1-x}|\square_2)Cl_2$.

9.314—A vacant site is represented by the single symbol □, without atomic symbol. The vacant position in lithium magnesium chloride can thus be shown as $Li_{2x}Mg_{1-x}\square_{1-x}Cl_2$ or $Li_{1-2y}Mg_y\square_yCl$.

Some authors find the symbols given in **9.312** and **9.313** unnecessary. We can write the spinel formula $Al^{III}{}_2oct\ Mg^{II}tet\ O_4$ and the inverse spinel type magnetite $(Fe^{II}Fe^{III})oct\ Fe^{III}tet\ O_4$. Maghemite ($\gamma$-$Fe_2O_3$) is $(Fe_{\frac{5}{3}}\square_{\frac{1}{3}})oct$ Fe tet O_4. Intermediates between magnetite and maghemite have been prepared; they can be written:

$$(Fe^{II}{}_{1-x}Fe^{III}{}_{1+2x/3}\square_{x/3})oct\ Fe^{III}tet\ O_4$$

A heated potassium chloride crystal with Schottky defects (cation vacancies *and* anion vacancies) can be written $(K_{1-\delta}\square_\delta)\ (Cl_{1-\delta}\square_\delta)$.

A silver bromide crystal with Frenkel defects (cation vacancies *and* interstitial cations, but with the anion lattice intact) is written $(Ag_{1-\delta}\square_\delta)\ (Ag_\delta|\triangle)Br$. Although the symbol |△ could be left out, its use improves clarity.

The α-modification of silver iodide, in which the cations are distributed at random over cation sites *and* interstitial sites, can be written $(Ag|\square, \triangle)I$.

9.32—In the discussion of catalytic reactions designation of a site in the surface by □surf is useful. An interstitial site in the surface will be Δsurf. An oxide ion in the surface of a metal oxide will then be designated by (O^{2-}| □surf). The type of surface site occupied is specified as in the example (O^{2-}|□oct, surf), where the oxide ion occupies a potentially octahedral site.

9.33—Electrons and positive holes bound in the field of an excess of positive or negative charge are designated by "e^-" and "v^+" respectively (v = vacancy)*.

Examples: Germanium doped with arsenic or gallium is $Ge_{1-\delta}As_\delta$ or $Ge_{1-\delta}Ga_\delta$ respectively, but if it is desired to emphasize the semiconductor properties this can be expressed by the formulae $Ge_{1-\delta}As^+{}_\delta e^-{}_\delta$ or $Ge_{1-\delta}Ga^-{}_\delta v^+{}_\delta$, although it is known that no more than 50% of the impurity atoms are ionized at room temperature.

Likewise sodium chloride with an excess of sodium has anion vacancies (F-centres) expressed by $Na^+Cl^-{}_{1-\delta}e^-{}_\delta$, or, if it is desired to show that the electron is trapped in an anion vacancy by $Na^+(Cl^-{}_{1-\delta}e^-{}_\delta$|□an).

Zinc oxide with an excess of zinc is on the contrary believed to contain interstitial cations (and electrons trapped by them), expressed by (Zn^{2+}|□) ($Zn^{2+}{}_\delta e^-{}_{2\delta}$|Δ)$O^{2-}$.

9.34—To indicate that two kinds of defect occur in association the symbol ≬ may be used.

Example:

Iron-deficient iron(II) oxide ($Fe^{II}{}_{1-3x}Fe^{III}{}_{2x}$|Δ≬□$cat_{3x}$)O.

**Note*: Although authors in the semiconductor field have used n and p for electrons and positive holes respectively, this practice is confusing since these letters are pre-empted to designate neutrons and protons respectively.

10. POLYMORPHISM

Minerals occurring in Nature with similar compositions have different names according to their crystal structures; thus, zinc blende, wurtzite; quartz, tridymite, and cristobalite; *etc.* Chemists and metallographers have designated polymorphic modifications with Greek letters or with Roman numerals (α-iron, ice-I, *etc.*). The method is similar to the use of trivial names, and is likely to continue to be of use in the future in cases where the existence of polymorphism is established, but not the structures underlying it. Regrettably there has been no consistent system, and some investigators have designated as α the form stable at ordinary temperatures, while others have used α for the form stable immediately below the melting point, and some have even changed an already established usage and renamed α-quartz as β-quartz, thereby causing confusion. If the α–β nomenclature is used for two substances A and B, difficulties are encountered when the binary system A–B is studied.

A rational system should be based upon crystal structure, and the designations α, β, γ, *etc.*, should be regarded as provisional, or as trivial names. The designations should be as short and understandable as possible, and convey a maximum of information to the reader. The rules suggested here have been framed as a basis for future work, and it is hoped that experience in their use may enable more specific rules to be formulated at a later date.

10.1—For chemical purposes (*i.e.*, when particular mineral occurrences are not under consideration) polymorphs should be indicated by adding the crystal system after the name or formula. For example, zinc sulfide(cub.), or ZnS(cub.), corresponds to zinc blende or sphalerite, and ZnS(hex.) to wurtzite. The Commission considers that these abbreviations might with advantage be standardized internationally:

cub. = cubic; c. = body centred; f. = face-centred
tetr. = tetragonal
o-rh. = orthorhombic
hex. = hexagonal
trig. = trigonal
mon. = monoclinic
tric. = triclinic

Slightly distorted lattices may be indicated by use of the *circa* sign, ≈. Thus, for example, a slightly distorted face-centred cubic lattice would be expressed as ≈f.cub.

10.2—Crystallographers may find it valuable to add the space-group; it is doubtful whether this system would commend itself to chemists where **10.1** is sufficient.

10.3—Simple, well-known structures may also be designated by giving the type-compound in italics in parentheses; but this system often breaks down as many structures are not referable to a type in this way. Thus, AuCd above 70° may be written as AuCd(cub.) or as AuCd(*CsCl-type*); but at low temperature only as AuCd(o-rh.), as its structure cannot be referred to a type.

11. BORON COMPOUNDS*

11.1. BORON HYDRIDES

11.11—The name of BH_3 is borane and it and higher boron hydrides are called boranes. The number of boron atoms in the molecule is indicated by a Greek numerical prefix (except that the Latin nona and undeca are used instead of ennea and hendeca to conform with hydrocarbon nomenclature. The prefix for twenty is spelled icosa to agree with common practice in geometry as opposed to the practice in organic hydrocarbon nomenclature).
The number of hydrogen atoms in the molecule is indicated by enclosing the appropriate Arabic numeral in parentheses directly following the name. Substituted boranes should retain the original numerical suffix.

1. B_2H_6 diborane(6) (*Figure 1*)
2. B_4H_{10} tetraborane(10) (*Figure 2*)
3. B_5H_9 pentaborane(9) (*Figure 3*)
4. B_5H_{11} pentaborane(11) (*Figure 4*)
5. B_6H_{10} hexaborane(10) (*Figure 5*)
6. B_9H_{15} nonaborane(15)
7. $B_{10}H_{14}$ decaborane(14) (*Figure 6*)
8. $B_{20}H_{16}$ icosaborane(16) (*Figure 7*)

The boron atoms are numbered according to certain conventions.* The numbering of some of the boron hydrides is seen from their structural and diagrammatic formulae shown in *Figures 1–7*.

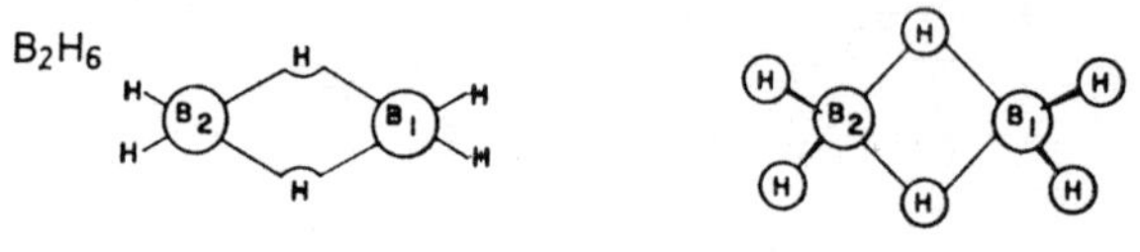

B_2H_6

Figure 1

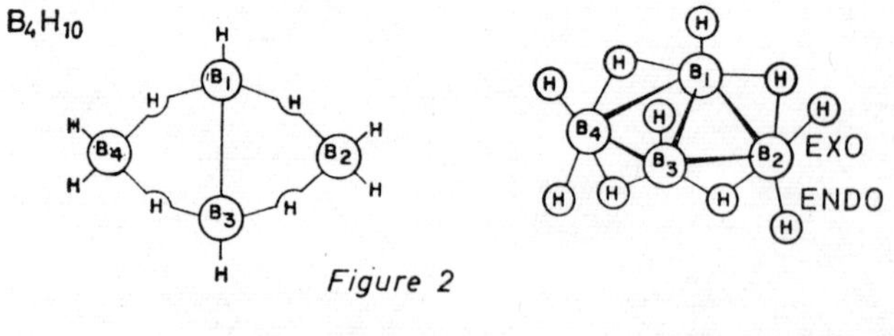

B_4H_{10}

EXO

ENDO

Figure 2

* For more detail concerning the nomenclature of inorganic boron compounds, reference should be made to *I.U.P.A.C. Information Bulletin*: Appendices on Tentative Nomenclature, Symbols, Units and Standards, No. 8 (1970) or *Inorg. Chem.*, 7, 1948 (1968).

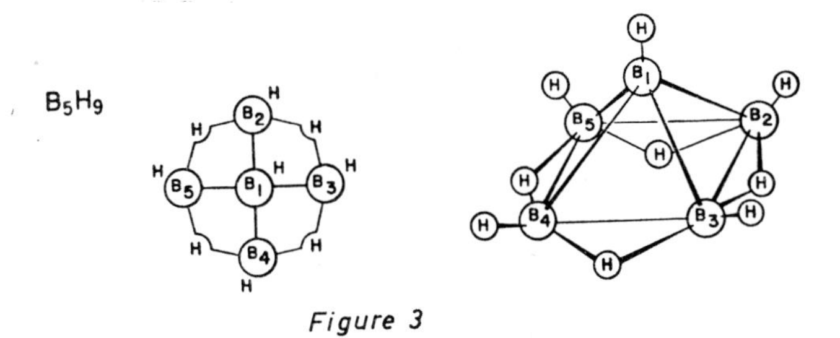

Figure 3

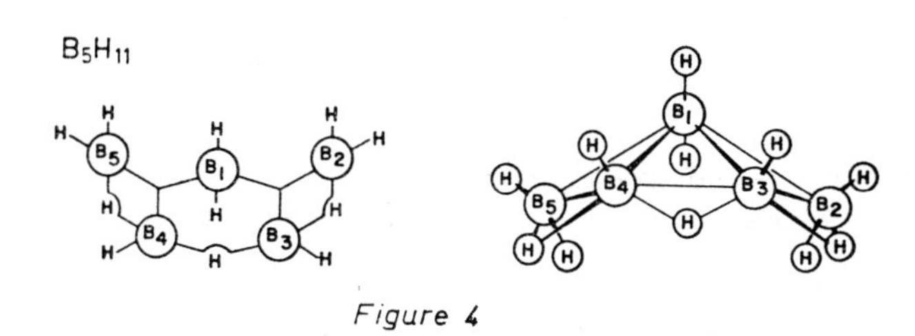

Figure 4

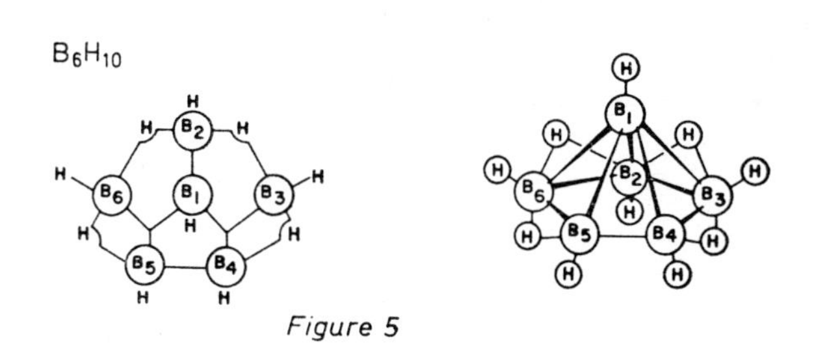

Figure 5

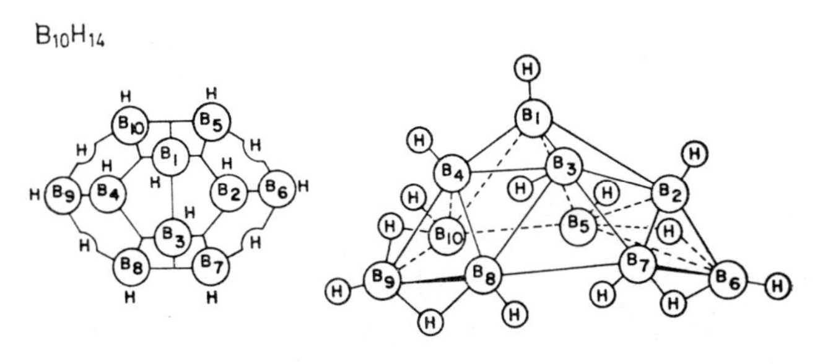

Figure 6

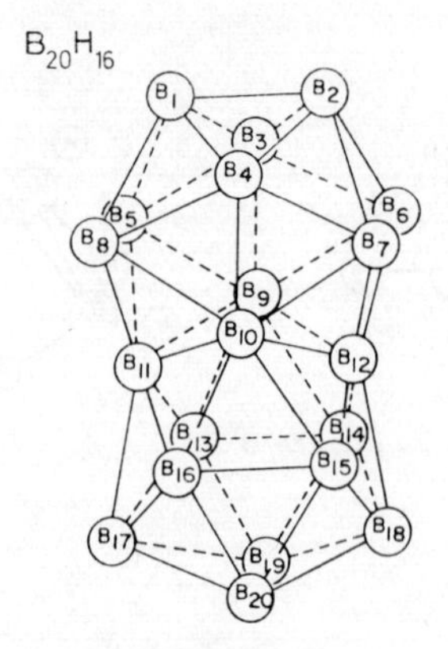

Figure 7

11.12—Prefixes *iso-* and *neo-* have been used to distinguish isomers of unknown structures. Once structures are known, a structural name is preferred.

Examples:

1. $B_{18}H_{22}$ decaborano(14)[6′,7′:5,6]decaborane(14) } (*Figure 8*)
2. *iso*-$B_{18}H_{22}$ decaborano(14)[6′,7′:6,7]decaborane(14) } (*Figure 8*)

The above names are fusion names formed in the same manner as for organic compounds.

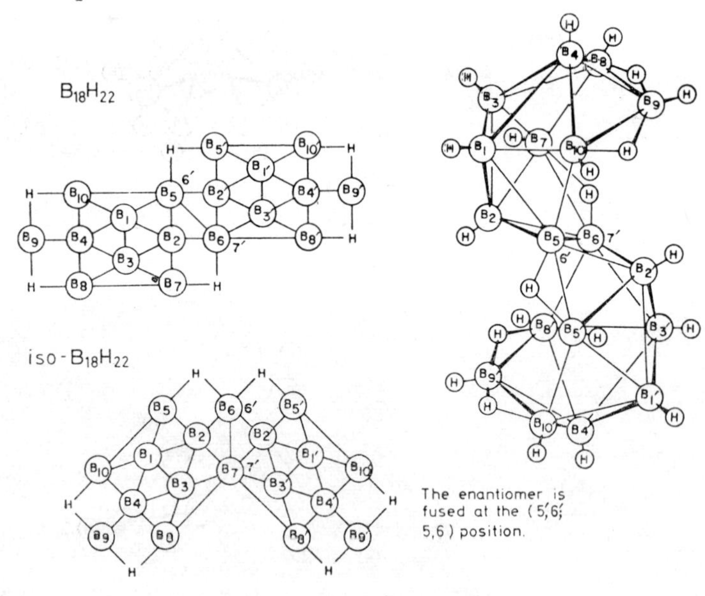

Figure 8

11.13—The polyboranes and their derivatives may be considered to consist of two general classes: (1) closed structures (*i.e.*, structures with boron skeletons that are polyhedra having all triangular faces) and (2) non-closed structures. The members of the first class are designated by the prefix *closo*. Some members of the second class have structures very close to a closed structure. These may be denoted by the prefix *nido* (from Latin nidus, nest).

Examples:

1. $B_{10}H_{14}$ *nido*-decaborane(14) (*Figure 6*)
2. $B_{20}H_{16}$ *closo*-icosaborane(16) (*Figure 7*)

11.2. BORANES WITH SKELETAL REPLACEMENT

11.21—The names of the general classes of compounds in which one or more boron atoms in the network have been replaced by another atom are formed by an adaptation of organic replacement nomenclature as carbaboranes, azaboranes, thiaboranes, *etc.* In this adaptation, a BH group is replaced by an EH group where E is the replacing atom (irrespective of its valency).

Example:

$B_{10}C_2H_{12}$ dicarbadodecaborane(12)

This compound which is very stable and has many derivatives, is named as the dicarba replacement derivative of the unknown $B_{12}H_{12}$. It may be derived from the stable ion $B_{12}H_{12}^{2-}$ (the numbering of which is shown in *Figure 9*) by replacing two B^- with the isoelectronic carbon atoms.

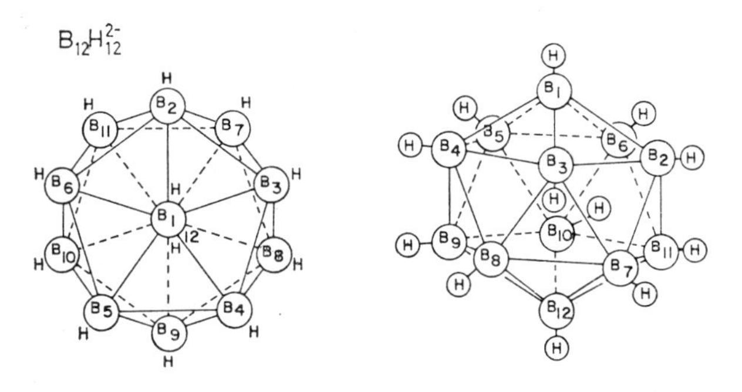

Dodecahydro-*closo*-dodecaborate(2-)
(icosahedron)
Figure 9

11.22—Numerical locants, and prefixes *closo-* or *nido-* are used when the structures of such compounds are known.

Examples:

1. $B_{10}C_2H_{12}$ 1,2-dicarba-*closo*-dodecaborane(12)
   1,7-dicarba-*closo*-dodecaborane(12)
   1,12-dicarba-*closo*-dodecaborane(12) } isomers
2. $B_{10}SH_{12}$ 7-thia-*nido*-undecaborane(12)

The names carborane and barene are not recommended.

11.3. BORON RADICALS

11.31—Radicals derived from borane, BH_3, are named as follows:

H_2B—	boryl (and *e.g.*, Cl_2B— dichloroboryl, $(HO)_2B$— dihydroxyboryl)
$HB<$	boranediyl
$B\leqslant$	boranetriyl

The name boryl rather than boranyl serves the purpose of avoiding confusion between diboryl (meaning two boryl groups) and diboranyl.

11.32—Radicals derived from diborane, B_2H_6, are named as follows:

H_2BH_2BH—	diboranyl
—HBH_2BH—	1,2-diboranediyl
$H_2BH_2B<$	1,1-diboranediyl
$H_2B(\mu\text{-}H)BH_2$ (bridged structure: H above, bonds below)	(1–2)diboranyl

It may be necessary for clarity to insert the numerical designation of the number of hydrogen atoms in the parent borane before the radical ending, *e.g.*, diboran(6)yl and diboran(4)yl.

11.33—The position of attachment of a radical shall be given the lowest possible designation and is indicated by placing the appropriate numeral or symbol before the radical name.

Example:

Structure: central B bonded (bond upward) with four HB/BH groups (HB, BH, HB, BH) bridged by H atoms (H top, H left, H right, H bottom) — 1-pentaboran(9)yl

11.4. SUBSTITUTION PRODUCTS OF BORANES

11.41—When it is impossible or unnecessary to give structural information in the name, the stoicheiometric name is used (**2.22** and **2.25**).

Examples:

1. B_4Cl_4 tetraboron tetrachloride
2. B_8Cl_8 octaboron octachloride

11.42—If their structures are known, substitution derivatives may be named by substitutive nomenclature as derivatives of real or postulated boranes. Replacement of a bridging hydrogen atom is indicated by the symbol μ. If it is necessary to distinguish between bridge positions, the bridge positions are indicated by designating the numbers of the boron atoms across which bridging occurs followed by a hyphen. Otherwise numerical locants are used. However, the hydrogen atoms of a BH_2 group are different and may be distinguished by the italicized prefixes *exo-* and *endo-* (see *Figure 2*).

Examples:

1. B_4Cl_4 — tetrachloro-*closo*-tetraborane(4)
2. H_2BH_2BHCl — 1-chlorodiborane(6)
3. 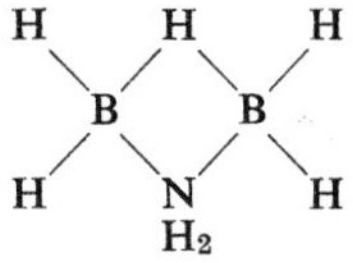
```
H     H     H
  \  /  \  /
   B     B
  /  \  /  \
H     N     H
      H2
```
μ-aminodiborane(6)

4. 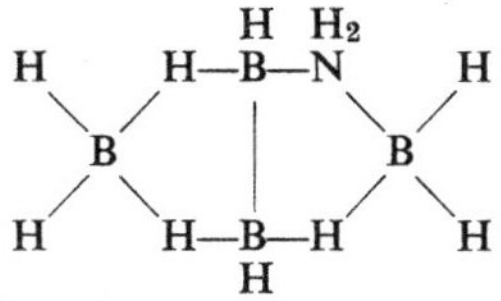
```
          H  H2
H      H—B—N       H
  \   /    |   \  /
   B       |    B
  /   \    |   /  \
H      H—B—H       H
          H
```
1,2-μ-aminotetraborane(10)

11.5. ANIONS DERIVED FROM THE BORANES

Anions derived from boranes or heteroboranes are named in accordance with **7.24** and **7.31**, except that "hydro" is used in preference to "hydrido".

Examples:

1. $[BH_4]^-$ — tetrahydroborate(1−)
2. $[B(CH_3)_2H_2]^-$ — dihydrodimethylborate(1−)
3. $[BCF_3F_3]^-$ — trifluoro(trifluoromethyl)borate(1−)
4. $[B_3H_8]^-$ — octahydrotriborate(1−)
5. $[B_{10}Cl_{10}]^{2-}$ — decachlorodecaborate(2−)
6. $[B_{10}H_{12}]^{2-}$ — dodecahydro-*nido*-decaborate(2−)
7. $[B_{12}H_{12}]^{2-}$ — dodecahydro-*closo*-dodecaborate(2−)
8. $[B_{10}C_2H_{12}]^{2-}$ — dodecahydro-1,2-dicarba-*closo*-dodecaborate(2−)

11.6. CATIONS DERIVED FROM THE BORANES

In accordance with **7.24**, cations are given no distinguishing termination (such as -onium).

Examples:

1. $[BH_2py_2]^+$ — dihydrobis(pyridine)boron(1+)
2. $[B_{10}H_7(NH_3)_3]^+$ — triammineheptahydrodecaboron(1+)

11.7. SALTS DERIVED FROM THE BORANES

Salts are named listing the cation(s) followed by the anion(s).

Example:

$[BH_2(NH_3)_2]\,[B_3H_8]$ — diamminedihydroboron octahydrotriborate

TABLE I

ELEMENTS

Name	*Symbol*	*Atomic number*	*Name*	*Symbol*	*Atomic number*
Actinium	Ac	89	Mercury	Hg	80
Aluminium	Al	13	Molybdenum	Mo	42
Americium	Am	95	Neodymium	Nd	60
Antimony	Sb	51	Neon	Ne	10
Argon	Ar	18	Neptunium	Np	93
Arsenic	As	33	Nickel	Ni	28
Astatine	At	85	Niobium	Nb	41
Barium	Ba	56	Nitrogen	N	7
Berkelium	Bk	97	Nobelium	No	102
Beryllium	Be	4	Osmium	Os	76
Bismuth	Bi	83	Oxygen	O	8
Boron	B	5	Palladium	Pd	46
Bromine	Br	35	Phosphorus	P	15
Cadmium	Cd	48	Platinum	Pt	78
Caesium	Cs	55	Plutonium	Pu	94
Calcium	Ca	20	Polonium	Po	84
Californium	Cf	98	Potassium	K	19
Carbon	C	6	Praseodymium	Pr	59
Cerium	Ce	58	Promethium	Pm	61
Chlorine	Cl	17	Protactinium	Pa	91
Chromium	Cr	24	Radium	Ra	88
Cobalt	Co	27	Radon	Rn	86
Copper (Cuprum)	Cu	29	Rhenium	Re	75
Curium	Cm	96	Rhodium	Rh	45
Dysprosium	Dy	66	Rubidium	Rb	37
Einsteinium	Es	99	Ruthenium	Ru	44
Erbium	Er	68	Samarium	Sm	62
Europium	Eu	63	Scandium	Sc	21
Fermium	Fm	100	Selenium	Se	34
Fluorine	F	9	Silicon	Si	14
Francium	Fr	87	Silver (Argentum)	Ag	47
Gadolinium	Gd	64	Sodium	Na	11
Gallium	Ga	31	Strontium	Sr	38
Germanium	Ge	32	Sulfur	S	16
Gold (Aurum)	Au	79	Tantalum	Ta	73
Hafnium	Hf	72	Technetium	Tc	43
Helium	He	2	Tellurium	Te	52
Holmium	Ho	67	Terbium	Tb	65
Hydrogen	H	1	Thallium	Tl	81
Indium	In	49	Thorium	Th	90
Iodine	I	53	Thulium	Tm	69
Iridium	Ir	77	Tin (Stannum)	Sn	50
Iron (Ferrum)	Fe	26	Titanium	Ti	22
Krypton	Kr	36	Tungsten (Wolfram)	W	74
Lanthanum	La	57	Uranium	U	92
Lawrencium	Lr	103	Vanadium	V	23
Lead (Plumbum)	Pb	82	Xenon	Xe	54
Lithium	Li	3	Ytterbium	Yb	70
Lutetium	Lu	71	Yttrium	Y	39
Magnesium	Mg	12	Zinc	Zn	30
Manganese	Mn	25	Zirconium	Zr	40
Mendelevium	Md	101			

TABLE II

NAMES FOR IONS AND RADICALS

(In inorganic chemistry substitutive names are seldom used, but the organic-chemical names are shown to draw attention to certain differences between organic and inorganic nomenclature.)

	Name				
Atom or group	*as uncharged atom, molecule or radical*	*as cation or cationic radical*	*as anion*	*as ligand*	*as prefix for substituent in organic compounds*
H	(mono)hydrogen	hydrogen	hydride	hydrido	
F	(mono)fluorine	fluorine	fluoride	fluoro	fluoro
OF	oxygen (mono)fluoride				fluorooxy, F—O—
Cl	(mono)chlorine	chlorine	chloride	chloro	chloro
ClO		chlorosyl	hypochlorite	hypochlorito	chlorosyl, O=Cl—
ClO_2	chlorine dioxide	chloryl	chlorite	chlorito	chloryl
ClO_3		perchloryl	chlorate	chlorato	perchloryl
ClO_4			perchlorate	perchlorato	
ClS		chlorosulfanyl			chlorothio, Cl—S—
ClF_2	chlorine difluoride		difluorochlorate(I)		difluorochloro
Br	(mono)bromine	bromine	bromide	bromo	bromo
I	(mono)iodine	iodine	iodide	iodo	iodo
IO		iodosyl	hypoiodite		iodosyl
IO_2		iodyl			iodyl
ICl_2			dichloroiodate(I)		dichloroiodo
O	(mono)oxygen		oxide	oxo	oxo, O= oxy, —O—; oxido, $—O^-$
O_2	dioxygen	dioxygen(1+), O_2^+	peroxide, O_2^{2-} hyperoxide, O_2^-	peroxo dioxygen	dioxy, —O—O—
O_3	trioxygen (ozone)		ozonide		trioxy, —O—O—O—
H_2O	water			aqua	oxonio, H_2O^+—
H_3O		oxonium			
HO	hydroxyl		hydroxide	hydroxo	hydroxy

Table II continued

Atom or group	Name: as uncharged atom, molecule or radical	as cation or cationic radical	as anion	as ligand	as prefix for substituent in organic compounds
HO_2	perhydroxyl		hydrogenperoxide	hydrogenperoxo	hydroperoxy
S	(mono)sulfur		sulfide	thio, sulfido	thio, —S—; sulfido, $—S^-$ thioxo, S=
HS	sulfhydryl		hydrogensulfide	mercapto	mercapto
S_2	disulfur	disulfur(1+)	disulfide	disulfido	dithio, —S—S—
SO	sulfur monoxide	sulfinyl (thionyl)			sulfinyl
SO_2	sulfur dioxide	sulfonyl (sulfuryl)	sulfoxylate	sulfur dioxide	sulfonyl
SO_3	sulfur trioxide		sulfite	sulfito	sulfonato, $—SO_3^-$
HSO_3			hydrogensulfite	hydrogensulfito	sulfo, $(HO)O_2S—$
H_2S	dihydrogen sulfide				sulfonio, $H_2S^+—$
H_3S		sulfonium			
S_2O_3			thiosulfate	thiosulfato	
SO_4			sulfate	sulfato	sulfonyldioxy, $—O—SO_2—O—$
Se	(mono)selenium		selenide	seleno	seleno, —Se— selenoxo, Se=
SeO		seleninyl	selenoxide		seleninyl
SeO_2	selenium dioxide	selenonyl			selenonyl
SeO_3	selenium trioxide		selenite	selenito	
SeO_4			selenate	selenato	
Te	(mono)tellurium		telluride	telluro	telluro
CrO_2	chromium dioxide	chromyl			
UO_2	uranium dioxide	uranyl			
N	(mono)nitrogen		nitride	nitrido	nitrilo, N≡
N_2	dinitrogen	dinitrogen(1+), N_2^+		dinitrogen	azo, —N=N—; azino, =N—N=; diazo, $=N_2$; diazonio, $—N_2^+$
N_3			azide	azido	azido
NH	aminylene	aminylene	imide	imido	imino

Table II continued

Atom or group	Name: as uncharged atom, molecule or radical	as cation or cationic radical	as anion	as ligand	as prefix for substituent in organic compounds
NH_2	aminyl	aminyl	amide	amido	amino
NH_3	ammonia			ammine	ammonio, H_3N^+—
NH_4		ammonium			
NH_2O			hydroxylamide	hydroxylamido-*O*	aminooxy, H_2NO—
				hydroxylamido-*N*	hydroxyamino, HONH—
N_2H_3	hydrazyl	hydrazyl	hydrazide	hydrazido	hydrazino
N_2H_4	hydrazine			hydrazine	
N_2H_5		hydrazinium(1+)		hydrazinium(1+)	
N_2H_6		hydrazinium(2+)			
NO	nitrogen oxide	nitrosyl		nitrosyl	nitroso
N_2O	dinitrogen oxide			dinitrogen oxide	azoxy
NO_2	nitrogen dioxide	nitryl	nitrite	nitro (nitrito-*N*)	nitro, —NO_2
				nitrito-*O*	nitrosooxy, —O—N=O
NS		thionitrosyl			
NO_3			nitrate	nitrato	
N_2O_2			hyponitrite	hyponitrito	
P	(mono)phosphorus		phosphide	phosphido	phosphinetriyl
H_2P			dihydrogenphosphide	dihydrogenphosphido	phosphino
PH_3	phosphine			phosphine	phosphonio, H_3P^+—
PH_4		phosphonium			
PO		phosphoryl			phosphoroso, OP—
					phosphoryl, OP $\leqslant$
PS		thiophosphoryl			thiophosphoryl
PH_2O_2			phosphinate	phosphinato	
PHO_3			phosphonate	phosphonato	
PO_4			phosphate	phosphato	
$P_2H_2O_5$			diphosphonate	diphosphonato	
P_2O_7			diphosphate	diphosphato	
AsO_4			arsenate	arsenato	

Table II continued

Atom or group	Name: as uncharged atom molecule or radical	as cation or cationic radical	as anion	as ligand	as prefix for substituent in organic compounds
CO	carbon monoxide	carbonyl		carbonyl	carbonyl
CS		thiocarbonyl		thiocarbonyl	thiocarbonyl
HO_2C	carboxyl			carboxyl	carboxy
CO_2	carbon dioxide			carbon dioxide	carboxylato
CS_2	carbon disulfide			carbon disulfide	dithiocarboxylato
$ClCO$	chloroformyl			chloroformyl	chloroformyl
H_2NCO	carbamoyl			carbamoyl	carbamoyl
H_2NCO_2			carbamate	carbamato	carbamoyloxy
CH_3O	methoxyl		methoxide or methanolate	methoxo or methanolato	methoxy
C_2H_5O	ethoxyl		ethoxide or ethanolate	ethoxo or ethanolato	ethoxy
CH_3S	methylsulfanyl		methanethiolate	methylthio or methanethiolato	methylthio
C_2H_5S	ethylsulfanyl		ethanethiolate	ethylthio or ethanethiolato	ethylthio
CN		cyanogen	cyanide	cyano	cyano, —CN
					isocyano, —NC
OCN			cyanate	cyanato	cyanato, —OCN
				isocyanato	isocyanato, —NCO
ONC			fulminate	fulminato	
SCN		thiocyanogen	thiocyanate	thiocyanato	thiocyanato, —SCN
				isothiocyanato	isothiocyanato, —NCS
$SeCN$			selenocyanate	selenocyanato	selenocyanato, —SeCN
				isoselenocyanato	isoselenocyanato, —NCSe
CO_3			carbonate	carbonato	carbonyldioxy, —O—CO—O—
HCO_3			hydrogencarbonate	hydrogencarbonato	
CH_3CO_2		acetoxyl	acetate	acetato	acetoxy
CH_3CO	acetyl	acetyl		acetyl	acetyl
C_2O_4			oxalate	oxalato	

TABLE III

PREFIXES OR AFFIXES USED IN INORGANIC NOMENCLATURE

Multiplying affixes (a)	mono, di, tri, tetra, penta, hexa, hepta, octa, nona (ennea), deca, undeca (hendeca), dodeca, *etc.*, used by direct joining without hyphens.
(b)	bis, tris, tetrakis, pentakis, *etc.*, used by direct joining without hyphens, but usually with enclosing marks around each whole expression to which the prefix applies.
Structural affixes	italicized and separated from the rest of the name by hyphens.
antiprismo	eight atoms bound into a rectangular antiprism.
asym	asymmetrical.
catena	a chain structure; often used to designate linear polymeric substances.
cis	two groups occupying adjacent positions; sometimes used in the sense of *fac*.
closo	a cage or closed structure, especially a boron skeleton that is a polyhedron having all triangular faces.
cyclo	a ring structure*.
dodecahedro	eight atoms bound into a dodecahedron with triangular faces (*Figure 8A*, p. 64).
fac	three groups occupying the corners of the same face of an octahedron.
hexahedro	eight atoms bound into a hexahedron (*e.g.*, cube).
hexaprismo	twelve atoms bound into a hexagonal prism.
icosahedro	twelve atoms bound into a triangular icosahedron.
mer	meridional; three groups on an octahedron in such a relationship that one is *cis* to the two others which are themselves *trans*.
nido	a nest-like structure, especially a boron skeleton that is very close to a closed or *closo* structure.
octahedro	six atoms bound into an octahedron.
pentaprismo	ten atoms bound into a pentagonal prism.
quadro	four atoms bound into a quadrangle (*e.g.*, square).
sym	symmetrical.
tetrahedro	four atoms bound into a tetrahedron.
trans	two groups directly across a central atom from each other; *i.e.*, in the polar positions on a sphere.
triangulo	three atoms bound into a triangle.
triprismo	six atoms bound into a triangular prism.
η (eta or hapto)	signifies that two or more contiguous atoms of a group are attached to a metal.
μ (mu)	signifies that the group so designated bridges two or more centres of co-ordination.
σ (sigma)	signifies that one atom of the group is attached to a metal.

* *Cyclo* here is used as a modifier indicating structure and hence is italicized. In organic nomenclature, cyclo is considered to be part of the parent name since it changes the molecular formula and therefore is not italicized.

TABLE IV
ELEMENT SEQUENCE

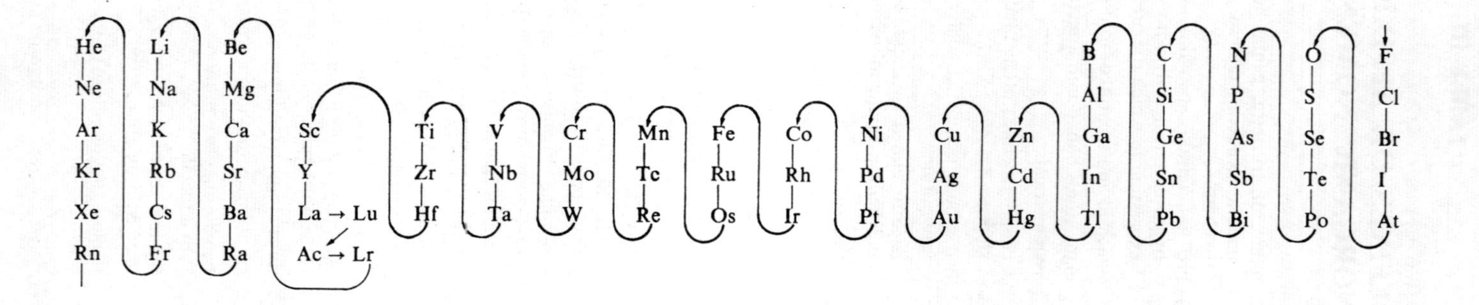

TABLE V

ELEMENT RADICAL NAMES

Element Name	*Radical Name**	*Element Name*	*Radical Name**
Actinium	Actinio	Mercury	Mercurio
Aluminium	Aluminio	Molybdenum	Molybdenio
Americium	Americio	Neodymium	Neodymio
Antimony	Antimonio	Neon	Neonio
Argon	Argonio	Neptunium	Neptunio
Arsenic	Arsenio	Nickel	Nickelio
Astatine	Astatio	Niobium	Niobio
Barium	Bario	Nitrogen	—
Berkelium	Berkelio	Nobelium	Nobelio
Beryllium	Beryllio	Osmium	Osmio
Bismuth	Bismuthio	Oxygen	—
Boron	Borio	Palladium	Palladio
Bromine	Bromio	Phosphorus	Phosphorio
Cadmium	Cadmio	Platinum	Platinio
Caesium	Caesio	Plutonium	Plutonio
Calcium	Calcio	Polonium	Polonio
Californium	Californio	Potassium	Potassio (Kalio)
Carbon	—	Praseodymium	Praseodymio
Cerium	Cerio	Promethium	Promethio
Chlorine	Chlorio	Protactinium	Protactinio
Chromium	Chromio	Radium	Radio
Cobalt	Cobaltio	Radon	Radonio
Copper (Cuprum)	Cuprio	Rhenium	Rhenio
Curium	Curio	Rhodium	Rhodio
Deuterium	Deuterio	Rubidium	Rubidio
Dysprosium	Dysprosio	Ruthenium	Ruthenio
Einsteinium	Einsteinio	Samarium	Samario
Erbium	Erbio	Scandium	Scandio
Europium	Europio	Selenium	Selenio
Fermium	Fermio	Silicon	Silicio
Fluorine	—	Silver (Argentum)	Argentio
Francium	Francio	Sodium	Sodio (Natrio)
Gadolinium	Gadolinio	Strontium	Strontio
Gallium	Gallio	Sulfur	Sulfurio(Thio)
Germanium	Germanio	Tantalum	Tantalio
Gold (Aurum)	Aurio	Technetium	Technetio
Hafnium	Hafnio	Tellurium	Tellurio
Helium	Helio	Terbium	Terbio
Holmium	Holmio	Thallium	Thallio
Hydrogen	—	Thorium	Thorio
Indium	Indio	Thulium	Thulio
Iodine	Iodio	Tin (Stannum)	Stannio
Iridium	Iridio	Titanium	Titanio
Iron (Ferrum)	Ferrio	Tritium	Tritio
Krypton	Kryptonio	Tungsten	(Tungstenio)
Lanthanum	Lanthanio	(Wolfram)	Wolframio
Lawrencium	Lawrencio	Uranium	Uranio
Lead (Plumbum)	Plumbio	Vanadium	Vanadio
Lithium	Lithio	Xenon	Xenonio
Lutetium	Lutetio	Ytterbium	Ytterbio
Magnesium	Magnesio	Yttrium	Yttrio
Manganese	Manganio	Zinc	Zincio
Mendelevium	Mendelevio	Zirconium	Zirconio

* In languages in which the ending "io" cannot be used another similar ending may be substituted.

APPENDIX

Collective names for Groups of Elements

If group names are needed they should be triels (B, Al, Ga, In and Tl), tetrels (C, Si, Ge, Sn and Pb) and pentels (N, P, As, Sb and Bi), with trielide, tetrelide and pentelide respectively for the binary compounds.

The use of other collective names such as "pnicogen" is not approved **(1.21)**.

INDEX

The numbers refer to Sections

Abbreviations for ligand names—7.35
Absolute configurations of coordination compounds—7.433, 7.8, 7.84
Acid anhydrides—5.32
Acid halogenides—5.31
"Acid" salts—6.2
Acidic hydrogen—6.323
Acids—5
Acids derived from polyatomic anions—5.2
Acids, formulae of—2.15, 2.163, 5
Actinoids—1.21
Addition compounds—8
Affixes used in inorganic nomenclature—Table III
Alcoholates—8
Alkali metals—1.21
Alkaline-earth metals—1.21
Allotropes—1.4
Alloys—*see* Intermetallic compounds
Amides—5.34
Ammine—7.322
"Ammoniates"—8
Anionic ligands—7.31
Anionic groups, abbreviations for—7.35
Anion names—3.2, 6.33, Table II
  monoatomic—3.21
  polyatomic—3.22
  of polyacids—4.12
Anions derived from boranes—11.5
Atomic number—1.31
Anticlinal (ac)—7.434
Antiperiplanar (ap)—7.434
antiprismo—Table III
Aqua, use of, in coordination complexes—7.322
Associated defects—9.34
asym—Table III
Azaboranes—11.21
"Basic" salts—6.4
Berthollides—9
Bidentate ligands—7.1
Binary acids—5.1
Binary compounds between non-metals—2.161
Boranes—11.11, 11.2
Boranes, anions derived from—11.5
Boranes, cations derived from—11.6
Boranes, radicals derived from—11.3
Boranes, salts derived from—11.7
Boranes, substitution products of—11.4
Boron compounds—11
Boron hydrides—11.1
Boron radicals—11.3
Brackets—0.32, 7.21
Bridged coordination compounds—7.61, 7.712
Bridging group—7.1, 7.61
Carbaboranes—11.21
Carbonyl, as ligand—7.323
catena—1.4, Table III
Cationic ligands—7.32
Cationic radicals—3.32
Cation names—3.1, 6.32, Table II
  monoatomic—3.11
  polyatomic—3.12 to 3.17
Cations derived from boranes—11.6
Central atom in complex compounds—2.24, 7.1, 7.21
Chain compounds, formulae of—2.162
Chain structure—1.4
Chalcogenides—1.21
Chalcogens—1.21
Chelate ligand—7.1
Characteristic atom in complex compounds—2.24
Chiral complexes—7.873
Chloroacids—5.24
cis—Table III
Clathrates—8
closo—Table III
Cluster—7.72

Complex, definition of—2.24
Complex compounds, formulae and names for—7.2
π-Complexes—7.4
Complexes with unsaturated groups—7.4
Complexes with unsaturated molecules—7.4
Condensed heteropolyanions—4.22
Configurational chirality—7.87
Conformation of chelate rings—7.8, 7.83
Conformation of ferrocene derivatives—7.434
Conformational chirality of a chelate ring—7.88
Coordination, centres of, direct linking between—7.71
Coordinating atoms—7.1
Coordination compounds—7
Coordination number—0.2
Coordination sites, designation of active—7.34
Crystalline phases of variable composition—9
Curoids—1.21
Cyclic heteropolyanions—4.213
Cyclic isopolyanions—4.14
Cyclopentadienyl complexes—7.43
cyclo—1.4, Table III
Defects, associated—9.34
Degree of polymerization—2.251
Designation of active coordination sites—7.34
Deuterium—1.11, 1.15
Diborane, radicals derived from—11.32
Different modes of linkage of ligands in coordination compounds—7.33
Dinuclear compounds with bridging groups—7.6
Dinuclear compounds without bridging groups—7.7
Direct linking between centres of coordination—7.71
dodecahedro—Table III
Donor-acceptor complexes—8
Double hydroxides—6.5
Double oxides—6.5
Double salts—6.3
Doubly-bonded ligand atoms—7.422
Electronegative constituent in names—2.22, 2.23
Electrons—9.33
Electropositive constituent in formulae—2.15
Electropositive constituents in names—2.21
Element radical names—Table V
Element sequence—Table IV
Elements—1, Table I
  atomic numbers of—Table I
  classification as: metals, semi-metals and non-metals—1.22
  groups of, names for—1.2
  new—1.13, 1.14
  symbols of—Table I
    in italic letters—0.34
Empirical formula—2.12
Enantiomers—7.433, 7.512
Enclosing marks—0.32, 7.21
Esters—5.33
Etherates—8
EWENS-BASSETT system—2.252
Extended structures—7.62
fac—Table III
Ferrocene—7.43, 7.431
Ferrocenyl radicals—7.432
Formulae—2, 2.1
Frenkel defects—9.314
Full characterization of absolute configuration of coordination compounds—7.86
Functional derivatives of acids—5.3
Geometrical isomerism—7.51
Groups of elements—1.2
Halides—1.21
Halogens—1.21
Hapto, η prefix, in structure designation—7.421
Heteropolyanions—4, 4.2
  chain or ring structure—4.21
  condensed—4.22
hexahedro—Table III
hexaprismo—Table III
Holes, positive—9.33
Homoatomic aggregates—7.72
Hydrates—8
Hydration of cations—6.322

Hydrocarbon-metal compounds, unsaturated—7.4
Hydrides—2.3
Hydrocarbon radicals—7.313
Hydroxide salts—6.4
icosahedro—Table III
Intermetallic compounds
  formulae—2.17
  names—2.22, 9
Interstitial compounds
  formulae—2.17
  names—9
Interstitial sites—9.31
Interstitial solutions—9
Ionic charge—1.31
Ions, names for—3, Table II
Isomers due to chirality (asymmetry) —7.52, 7.8
Isomers, designation of, in coordination compounds—7.5
Isomorphous replacement—9
Isopolyanions—4, 4.1
Isotopes, naming of—1.15
Isotopically labelled compounds—1.32
Italic letters—0.34
Lanthanoids—1.21
Lattice compounds—8
Ligancy—7.1
Ligands—2.24, 7.1
Ligands, different modes of linkage—7.33
Ligands, anionic—7.31
Ligands, cationic—7.32
Ligands, names for—7.3
Ligands, neutral—7.32
Ligands, order of citation of—7.25
Ligands with central branching—7.513
Ligating atoms—7.1
Locant designators, assignment of, in coordination compounds—7.512, 7.514, 7.613, 7.614
Locant designators for chelate ligands—7.513
Mass number—1.31
mer—Table III
Metallocenes—7.43
"Metalloid"—1.22
Metal-metal bonds—7.711
"Mixed hydroxides"—6.5
"Mixed oxides"—6.5
Molecular compounds—8
Molecular formula—2.13
Multidentate ligand—7.1
Multiplicative numerals, multiplying affixes and multiplying prefixes—0.31, 2.251, 7.21, Table III
Neutral groups, abbreviations for—7.35
Neutral ligands—7.32
Neutral radicals—3.32
nido—Table III
Nitriles—5.35
Nitrilo—3.33
Nitrogen cations—3.15
Nitrosyl—7.323
Noble gases—1.21
Nonstoicheiometric compounds (berthollides)—9
Nuclear or central atom—7.1
Nuclear reaction, equation for—1.31
Numbers, use of, in naming inorganic compounds—0.33
Numerical prefixes—2.251
Order of citation of ligands—7.25
"Ocene" names—7.43
octahedro—Table III
Optically active ligands—7.85
Oxidation number—0.1, 2.252, 7.22
  use of Roman numerals for indicating—0.33
Oxidation state—*see* Oxidation number
Oxide salts—6.4
Oxoacids—5.21
  list of—5.214
Oxonium ion—3.14
pentaprismo—Table III
Peroxoacids—5.22
"Pnicogen"—1.21
"Pnictides"—1.21
Polyatomic anions—3.22
Polyatomic cations—3.13
Polyboranes—11.13
Polymers—1.4, 4, 7.62
Polymorphism—10
Polynuclear compounds with bridging groups—7.6
Polynuclear compounds without bridging groups—7.7
Positive holes—9.33

Prefixes used in inorganic nomenclature —Table III
Preamble—0
Prefixes list—Table III
  hydroxy—6.42
  hypo—3.224, 5.211, 5.214
  per—3.224, 5.212, 5.214
  ortho—5.213
  oxy—6.42
  meta—5.213
  peroxo—5.22
  pyro—5.213
Prefix, numerical, in indexes—2.251
Proportions of the constituents—2.25
Protium—1.15
Pseudobinary acids—5.1
quadro—Table III
Radicals—3.3
  boron containing—11.3
  list of names for—Tables II, V
Ring structure—1.4
Salts—6
Salts containing acid hydrogen—6.2
Salts derived from the boranes—11.7
"Sandwich compounds"—7.43
Schottky defects—9.314
Semiconductors—9.33
Sequence-rule method—7.433
Simple salts—6.1
Site in the structure—9.311
  interstitial—9.311
  in the surface—9.32
  vacant—9.314
Six-coordinated complexes, absolute configurations concerned with—7.8
Skeletal replacement—11.2
Solvates—8
STOCK system—0.1, 2.252
Stoicheiometric composition—2.25, 7.41
Stoicheiometric prefixes—2.251, 7.22
Structural formula—2.14
Structural affixes—Table III
Structural prefixes—2.19, 7.23, Table III
Structure, designation of, in complexes —2.19, 7.42
Sub-groups of the elements—1.21
Substitution products of boranes—11.4
"Substitutive names"—3.33
sym—Table III
Symmetrical branched ligands—7.513
Symmetrical linear ligands—7.513
Synclinal (sc)—7.434
Systematic names—2.2
Terminations—7.24
  -ate—2.23, 3.223
  -ic—2.2531, 5.21
  -ide—2.23, 3.22
  -ite—2.23, 3.22, 3.224
  -o—2.24, 7.311
  -ous—2.2531, 5.21
tetrahedro—Table III
Thiaboranes—11.21
Thioacids—5.23
trans—Table III
Transition elements—1.21
triangulo—Table III
Triple, etc., salts—6.3
triprismo—Table III
Tritium—1.11, 1.15
Trivial names—2.4, 5.214
Unidentate ligand—7.1
Unsymmetrical branched ligands—7.513
Unsymmetrical linear ligands—7.513
Uranoids—1.21
Vacant sites—9.31